Introduction to Molecular Cloning Techniques

Introduction to Molecular Cloning Techniques

by

Gérard Lucotte and François Baneyx

Gérard Lucotte
Laboratoire d'Anthologie
  Moléculaire
CHU de Cochin-Port Royale
24, rue du Fb. St. Jacques
75004 Paris
France

François Baneyx
Department of Chemical Engineering
University of Washington
Seattle, WA 98195

This book is printed on acid-free paper.

Library of Congress Cataloging-in-Publication Data
Lucotte, Gérard
   [Techniques de clonage moléculaire.   English]
   Introduction to molecular cloning techniques / by Gérard Lucotte and François Baneyx.
     p.   cm.
   Includes bibliographical references and index.
   ISBN 1-56081-613-9.   ISBN 3-527-89613-9 (Weinheim)
   1. Molecular cloning—Laboratory manuals. I. Baneyx, François. II. Title
QH442.2L8313   1993
574.87′328′0724—dc20   93-22418
                                                   CIP

Originally published as "Techniques de clonage moléculaire" by Technique et Documentation
   Lavoisier, Paris, 11, rue Lavoisier, 75008 Paris, France and Éditions, Médicales
   Internationales, Allée de la Croix Bossée, 94234 Cachan Cedex, France

© 1993 VCH Publishers, Inc.

ISBN 1-56081-613-9 VCH Publishers
ISBN 3-527-89613-9 VCH Verlagsgesellschaft

Printing History
10 9 8 7 6 5 4 3 2 1

Published jointly by:

VCH Publishers, Inc.
220 East 23rd Street
New York, New York
10010-4606

VCH Verlagsgesellschaft mbH
P.O. Box 10 11 61
D-69451 Weinheim
Federal Republic of Germany

VCH Publishers (UK) Ltd.
8 Wellington Court
Cambridge CB1 1HZ
United Kingdom

Preface

This book offers a simplified summary of the basic principles and techiques of genetic engineering. It is focused entirely on the most widely used host, the gram-negative bacterium *Escherichia coli*. In addition to extensive descriptions of cloning vectors and essential recombinant DNA methodologies, the steps involved in the construction of genomic, cDNA and cosmid libraries are discussed. The different chapters tackle important aspects of molecular cloning by providing the necessary biochemistry and microbiology background in an effort to clearly introduce the pertinent genetic engineering concepts. Examples of routinely used experimental protocols and solved problems are also provided at the end of each chapter in order to extend their theoretical content and familiarize the reader with laboratory techniques. The text relies heavily on more than ten years of teaching experience by Dr. Lucotte in the biology, pharmaceutical and medical fields and contains a number of original experimental protocols.

G. Lucotte
F. Baneyx

April, 1993

Contents

PART III Cloning Vectors   45

CHAPTER 5. Bacteriophage λ   47

CHAPTER 6. Bacteriophage Cloning Vectors   53

CHAPTER 16. Enriching for Specific mRNA Molecules   133

CHAPTER 17. Complementary DNA Synthesis   139

PART V Cloning Techniques   151

CHAPTER 18. Ligation   153

Part I
The Host

1

Escherichia coli: A Host for Genetic Engineering

At present, the most widely used host in genetic engineering is the gram-negative bacterium *Escherichia coli.* The choice of this system was motivated by the fact that *E. coli* genetics are well understood. In addition, this bacterium can be grown to high density on inexpensive substrates. Most of the routinely used *E. coli* laboratory strains are derived from a parental strain named *K-12.* A very large number of *K-12* mutants are currently available and have proved invaluable for certain genetic engineering applications. (See Bachman, 1972, for the genealogy, origin, and order of isolation of a number of mutants.)

In spite of the progress accomplished in the characterization and use of other hosts (e.g., yeasts, bacillus, pseudomonas, plant, insect, mammalian cells . . .), *E. coli* often remains an obligated shuttle host in which most recombinant DNA manipulations are accomplished. Appendix 1 gives a short list of *E. coli* strains that have proven useful for recombinant DNA applications. Some important *E. coli* characteristics are described in this chapter.

1.1. Structure, Growth Conditions, Genetic Markers, and Nomenclature

E. coli is a rod-shaped bacterium surrounded by a cell envelope consisting of an *outer* and an *inner membrane* formed by closely associated phospholipid bilayers. The outer membrane is itself formed by a protein-phospholipid region and a *peptidoglycan* layer. The latter is a rigid polymer of polysaccharides cross-linked by a pentaglycine peptide that closes on itself to surround the entire inner membrane and cytoplasm of the bacterium. The gel-like space between the peptidoglycan layer and the inner (also known as plasma or cytoplasmic) membrane is called the *periplasmic space.* This cellular compartment contains proteases, nucleases, and proteins that bind and transport specific ions, sugars, and amino acids. The inner membrane provides the major osmotic barrier and performs an essential role by determining which molecules can enter or leave the cytoplasm. The latter contains a single, supercoiled, double-stranded chromosome about 1,300 μm in length. (Compare to the bacterium length and diameter, which are 3 μm and 1 μm, respectively!)

All laboratory *E. coli* strains are able to grow on a chemically well-defined substrate, the *minimal medium,* which contains, in addition to the organic compounds necessary for the cell to harvest energy and carbon atoms, trace amounts of inorganic compounds and ions. A typical minimal medium is *M9.* It is characterized by a high phosphate content that acts both as a source of this element and as a buffer.

In many instances, *E. coli* laboratory strains contain specific mutations that make their growth on minimal medium dependent upon the addition of various compounds (e.g., amino acids, purines, pyrimidines, vitamins . . .). Such mutated bacteria are known as *auxotrophs.* An inexpensive alternative to suplementing minimal medium with required additives consists in using *rich media* such as *LB.* The chemical composition of these media is not accurately known since they are prepared by hydrolysis of meat, milk, or yeast. For optimal growth, bacteria also require the presence of a *carbon source,* which is generally provided in the form of a sugar. Most strains grow on glucose; however, fructose, maltose, lactose, galactose, and glycerol can also be used as a carbon source. In formulating a medium for bacterial growth, one must carefully examine the genetic composition of the organism of interest (the *genotype*) since certain mutations prevent the cell from utilizing specific compounds (e.g., carbon sources).

Every mutation in the cell chromosome can be viewed as a *genetic marker;* their collection accurately describes a given strain. The nomenclature used to designate mutations was established by Demerec et al. (1966). The first three letters, in italics and lowercase, refer to the mutated locus (e.g., *his* for the histidine gene and *gal* for the galactose gene). The following letter, in capitals, indicates the polypeptide chain of interest or the regulatory element (operator or promoter) at the described locus (e.g., *his*A or *his*B). Finally, a number defines the particular mutation described (e.g., *his*A38).

Some strains harbor prophages in their genome and are referred to as *lysogenic* strains or *lysogens.* (See chapter 5.) The presence of such genetic elements is also indicated in the genotype. For instance, *nrd*A *his*C (λc1857) indicates that the corresponding bacterial strain is mutant in *nrd*A and *his*C and contains the λ prophage mutated for *c*1857. When the phage is a transduced particle, it may contain an *E. coli* gene that replaces one of its own essential genes. Such phages are known as *defective* and denoted by the prefix "d" preceding the denomination for the inserted bacterial gene (e.g., λd*lac*$^+$ is a bacteriophage λ

in which the *E. coli lac* gene replaces an essential phage gene). If, however, the gene that has been replaced in the phage is not essential, the transduced phage is still able to properly function (in other words, able to form plaques) and is denoted by the prefix "p" preceding the denomination for the inserted bacterial gene (e.g., λp*lac*$^+$).

1.2. Resistance to Antibiotics

Antibiotics are compounds that can act in a *bacteriostatic* or *bacteriocidal* fashion. In the first case they inhibit bacterial growth; in the second, they kill or lyse bacteria. For instance, at small doses, the antibiotic tetracyclin inhibits growth, but it acts as a bacteriocidal agent at high concentration. Certain plasmid-encoded genes can nullify the action of antibiotics. The mode of action of several important antibiotics and the way resistance can be conferred are discussed below.

- *Colicin E1* (colE1) causes lethal modifications in bacterial membranes. The colicin E1 resistance gene *(cea)* encodes a protein that interferes in an unknown fashion with colicin E1 to prevent its action.
- *Ampicillin* (Ap) is a penicillin derivative that kills growing bacteria by interfering with a late reaction step in bacterial cell-wall synthesis. Resistance to ampicillin is provided by the periplasmic enzyme β-lactamase (encoded by gene *bla*), which cleaves the β-lactam ring of ampicillin.
- *Tetracyclin* (Tc) interferes with bacterial protein synthesis at high concentration by associating with the 30S ribosomal subunit. The gene for tetracyclin resistance *(tet)* codes for a protein which modifies the cell wall and prevents transport of the antibiotic in the cytoplasm of the cell.
- *Chloramphenicol* (Cm) is a bacteriostatic agent that affects bacterial protein synthesis by associating with the 50S subunit of ribosomes, thereby preventing the formation of peptide bonds. Resistance to chloramphenicol is provided by the enzyme chloramphenicol acetyltransferase (encoded by *cat*), which acetylates and thereby inactivates the antibiotic.
- *Kanamycin* (Km) becomes bacteriocidal by binding to 70S ribosomes and inducing erroneous reading of mRNA. The gene *kan* codes for an enzyme that modifies the antibiotic and prevents its interaction with ribosomes.
- *Streptomycin* (Sm) is also a bacteriocidal agent that leads to errors in mRNA reading by binding to the 30S subunit of ribosomes. Resistance to streptomycin (gene *str*) is provided by an enzyme that acts similarly to the *kan* gene product for kanamycin.

1.3. Bacterial Growth

Under optimal conditions, the number of bacteria in a culture doubles about once every 20 min as a result of binary fission. When a given volume of growth

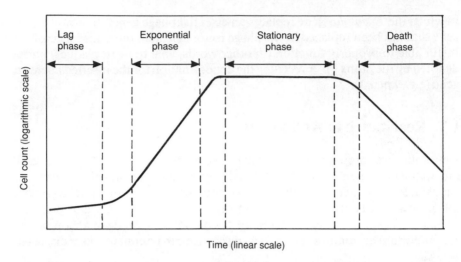

Figure 1.1. *Growth phases of an E. coli culture in liquid medium* The cell concentration is determined by measuring the spectrophotometric absorbance at 600 nm. There is a linear relationship between the logarithm of the absorbance and the incubation time in the exponential growth phase. (Under these conditions 1 OD_{600} = 8 × 10⁸ cells/ml.)

medium is supplemented with a small amount of bacterial cell (known as the *inoculum*), four different phases can be distinguished (Fig. 1.1):

- The *lag phase,* in which no growth occurs, is explained by the fact that the inoculated bacteria need a certain acclimation time before they start dividing.
- The *exponential phase* corresponds to the stage at which cells multiply and double their concentration at a constant rate. At 37°C, this stage usually lasts from 10 to 60 min. Most of the experiments performed in genetic engineering are carried out in this particular growth phase, which is sometimes referred to as the *logarithmic phase.* The main advantage of exponentially growing cells lies in the fact that they can be diluted in fresh medium of identical composition without decreasing the bacterial growth rate. As a consequence an *E. coli* culture can be propagated in exponential phase for an indefinite amount of time. The growth rate depends on a number of parameters, which include medium composition, strain genotype, growth temperature, and aeration rate of the culture.
- When medium composition changes, or when dissolved oxygen concentration becomes limiting, growth decreases and stops. At this point, the culture enters the *stationary phase,* where cells die as fast as they divide.
- At the end of this (usually long) plateau phase, the number of viable cells steadily diminishes as the culture enters the *death phase.*

1.4. Host Restriction and Modification

All bacterial strains derived from *E. coli K-12* normally produce a restriction enzyme *(EcoK)* that cleaves foreign DNA, as well as a methylase that protects their own DNA from self-digestion. (See chapter 2.) Because genetic engineering requires the transfer of extraneous DNA sequences in bacterial cells, it is necessary to bypass this system. This is usually achieved by using strains that have a defective restriction system (r_k^-). Most of the routinely used strains *(hsd$^-$)* are mutated both in their restriction and methylation systems $(r_k^- m_k^-)$.

1.5. Host Recombination

E. coli K-12 can achieve *homologous recombination* through its *rec*A system (the *rec*F and *rec*BCD gene products depend upon RecA) and its *rec*E system. Since the instability of some recombinant phages can be at least partially due to the host recombination systems, *rec*-strains are usually used in genetic-engineering applications. Nevertheless, such strains have the disadvantage of yielding a smaller amount of phages.

EXPERIMENTS

1.1. Growth Media

M9 Composition

Dissolve 6 g of $Na_2HOP_4 \cdot 2H_2O$, 3 g KH_2PO_4, 0.5 g NaCl, 1 g NH_4Cl in 1 L of ddH_2O. Adjust the pH to 7.4, autoclave, and transfer to a 45°C water bath. Add 2 ml of autoclaved 1 M $MgSO_4$ and 100 μl of 1 M $CaCl_2$. For plates, add 15 g of agar to the medium before autoclaving. The medium can be supplemented with 0.2% casamino acids, 0.2% glucose, and the appropriate antibiotics once it has cooled to 45°C.

LB (Luria-Bertani) Broth Composition

Dissolve 10 g of *Bacto*-tryptone, 10 g of NaCl, and 5 g of yeast extract in 1 L of ddH_2O, adjust the pH to 7.5, and cool to 45°C before adding 0.2% glucose and the appropriate antibiotics, if desired. For plates, dissolve 15 g of agar before autoclaving.

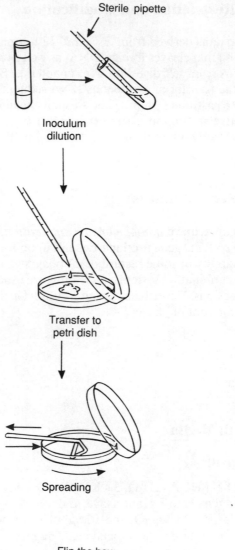

Sterile pipette

Inoculum
dilution

Transfer to
petri dish

Spreading

Flip the box
after 5 min

Incubate
at 37°C

Figure 1.2.   *Dilution and spreading of a bacterial culture on petri dishes.*

1.2. Spreading

In exponential phase or early stationary phase all bacteria are able to form isolated *colonies* on petri dishes. In order to get an easy count on plates, the number of colonies should not be greater than 100. Since the normal cell concentration in a growing culture is about 10^7 cells/mL, it is necessary to dilute them 10 to 100-fold prior to spreading on a plate. Fig. 1.2 shows how spreading should be performed. The characteristic aspect of bacterial colonies on an agar plate is shown in Fig. 1.3.

1.3. Isolation of a Single Colony

In order to eliminate all the possibilities of heterogeneity and contamination, bacterial strains should be grown from a single colony. This is experimentally achieved by picking a single colony at the surface of a plate. A flame-sterilized loop is used. It is cooled by dipping it into the agar. The selected colony is then transferred to a sterile tube containing 1 ml of growth medium, vortexed for a few seconds to disperse the bacteria, and incubated overnight in a constant-temperature water bath. The next day, a flame-sterilized loop is dipped into the culture and used to make a primary streak of the inoculum on a fresh plate.

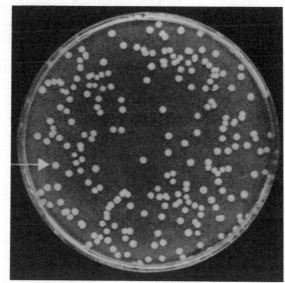

A bacterial colony

Figure 1.3. *Bacterial colonies on an agar plate.* A large number of colonies (about 2 mm in diameter) are visible on petri dishes containing a solid layer of nutrient agar after spreading and 24 h incubation at 37 °C.

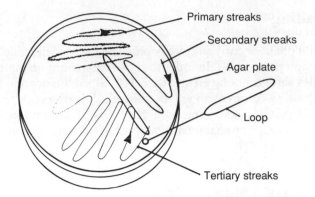

Figure 1.4. *Isolation of a single uncontaminated bacterial colony.*

Using the loop, secondary and tertiary streaks are generated from the primary as illustrated in Fig. 1.4. Once covered, the plates are inverted and incubated overnight at 37°C. This operation yields individual colonies at the end of the tertiary streak that correspond to an uncontaminated strain (Fig. 1.5).

1.4. Use of Antibiotics

Streptomycin sulfate is prepared as a 20 mg/ml stock solution kept at −20°C. It is used in plates or liquid cultures at a concentration of 25 μg/ml.

Ampicillin stock solutions are at 25–100 mg/ml and kept frozen at −20°C. For plates, the working concentration is 50–100 μg/ml.

Tetracyclin is prepared in 50% ethanol at a concentration of 12.5 mg/ml,

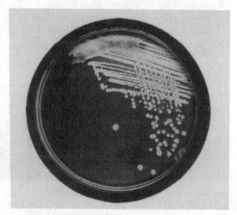

Figure 1.5. *After overnight incubation, single colonies are clearly visible at the end of the tertiary streaks generated as in Fig. 1.4.*

kept frozen, and used in plates at a concentration of 15–25 µg/ml with a medium that does not contain magnesium salts.

Chloramphenicol stock solutions are made at 34 mg/ml in pure ethanol. They are stored at $-20°C$ and used at 10–25 µg/ml.

Kanamycin stocks at 25 mg/ml are prepared in distilled deionized water (ddH$_2$O), stored frozen, and used at a 50 µg/ml.

1.5. Storage of Bacterial Strains

It is possible to keep bacterial strains for a few weeks directly on petri dishes sealed with Parafilm at 4°C. Medium-term storage is achieved by stabbing small bottles containing 2–3 ml of plate medium with a loop dipped into the bacterial culture of interest. These high-density *stabs* are stable at 4°C in the dark for about 1 year. They also constitute the most convenient way to ship out strains. For long-term storage, bacterial cultures are supplemented with glycerol to a final concentration of 15%, transferred to cryogenic tubes, and kept in a $-80°C$ cryogenic unit.

PROBLEMS

1.1 Strain Phenotype

The *phenotype* of an organism is defined as the collection of its observable properties. Strains 1–4 were streaked on sectors of plates containing either ampicillin or tetracyclin or chloramphenicol. In Fig. 1.6, a + sign indicates that the cells grew whereas no growth was observed in the case of a − sign. Give the phenotype of each of the four strains.

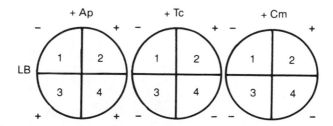

Figure 1.6.　Four different *E. coli* strains, labeled 1–4, have been streaked on different sectors of three petri dishes containing nutrient agar and either ampicillin or tetracyclin or cloramphenicol. Growing cells are labeled with a + sign, as are nongrowing cells with a − sign.

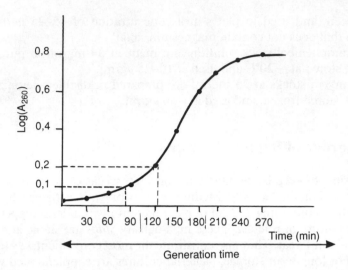

Figure 1.7. Typical growth curve for a bacterial culture. The logarithm of the absorbance at 600 nm is plotted vs. the incubation time.

1.2 Characteristics of the Exponential Phase

Suggest a way to measure the growth of bacteria in liquid medium. What is the main characteristic of cells growing in exponential phase? Fig. 1.7 shows a bacterial growth curve. How do you determine the doubling time and what is its value?

Part II
Restriction Enzymes, Methylation, Electrophoresis Techniques

2

Restriction Enzymes

2.1. Host Restriction and Modification

Restriction enzymes were discovered during the study of the restriction and modification of certain viruses, such as bacteriophage λ, by the host cell (Fig. 2.1). If a given stock of phage λ is able to grow on *E. coli,* the efficiency at which it infects the strain can be assessed by counting the number of lysed cells "holes" *(plaques)* it forms on a confluent bacterial lawn. (See chapter 5, section 3.) The *efficiency of plating* (EOP) is defined as the fraction of phage particles that can form a plaque. Thus, a phage stock for which each phage particle can infect a cell will have an EOP of 1. If a λ phage stock able to infect *E. coli* B with an EOP of 1 (noted λ^B in Fig. 2.1) is used to similarly infect a different *E. coli* strain (e.g., *E. coli* K), a very low EOP is obtained (about 10^{-4}, see Fig. 2.1). The phage λ^B is said to have been *restricted* by *E. coli* K.

However, if one of the few phages that have managed to infect *E. coli* K is used to reinfect the same strain, an EOP of 1 is obtained. These rare phages are said to have been *modified* by *E. coli* K (noted λ^K in Fig. 2.1). Such λ^K phages are, however, unable to reinfect *E. coli* B with a high EOP; in other words, they are now restricted by *E. coli* B.

These results can be explained as follows. *E. coli* K contains a *restriction endonuclease* (*Eco*K) that cuts DNA at very specific sequences (Dussoix and Arber, 1962). To prevent its own DNA from being destroyed, the strain also synthesizes a *methylase* (*Eco*K methylase) that methylates a residue in the DNA sequences recognized by the *Eco*K restriction enzyme. This modification efficiently prevents *E. coli* K DNA from being cleaved. When λ^B DNA, which

15

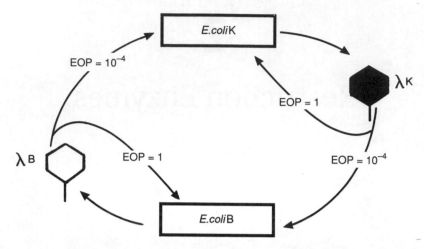

Figure 2.1. *Restriction-modification system of E. coli on bacteriophage* λ. When a stock of λ^B is used to infect *E. coli* B, its EOP is 1. Infection is markedly restricted when the same stock is used to infect *E. coli* K. The few λ^K phages that have managed to grow on *E. coli* K can be used to reinfect the same strain at an EOP of 1. (See text for details.)

contains the DNA sequences recognized by the *Eco*K endonuclease, is used to infect *E. coli* K, it is recognized as "foreign" since these sequences have not been modified by the *Eco*K methylase. The phage DNA is consequently destroyed through the action of the *Eco*K endonuclease. However, a small number of phages become methylated and therefore resistant to digestion by *Eco*K. Such phages (λ^K) are replicated in the presence of the *Eco*K methylase, and thus are no longer recognized as foreign when used to reinfect *E. coli* K. As a consequence, they score a high EOP on this strain. Since *E. coli* B also contains an *Eco*B restriction endonuclease and an *Eco*B methylase that function similarly but recognize different DNA sequences than the *Eco*K system, most λ^K phages will be destroyed upon infection of *E. coli* B.

In *E. coli,* the restriction–modification system involves the action of three linked genes: *hsd*R, *hsd*M, and *hsd*S; *hsd*R encodes the restriction endonuclease, while the *hsd*M gene product methylates DNA to protect it from cleavage by the endonuclease. The *hsd*S gene product does not display a restriction or modification activity *per se* but is essential for both enzymes to recognize the specific DNA sequences to be methylated or cleaved.

E. coli K and B restriction endonucleases have been studied in detail (Meselson and Yuan, 1968) and display similar properties. Both are multimeric enzymes that are, in fact, able to both modify and cleave DNA. Their activity is dependent upon the presence of magnesium and the cofactors ATP and *S*-adenosylmethionine. These enzymes have peculiar cutting properties. They recognize specific sequences on double-stranded DNA, move 1,000–5,000 nucleotides away from these sites, and cut (apparently randomly) a single DNA strand to liberate oligonucleotides of about 75 bp in length.

Enzymes presenting such properties are known as *type I* restriction endonucleases. Because of their lack of cleavage specificity, they are not used in genetic engineering.

2.2. Isolation of Type II Restriction Endonucleases

A classic example of what is now known as *type II* restriction endonucleases was discovered in *Haemophilus influenzae* Rd in 1970 (Smith and Wilcox, 1970; Kelly and Smith, 1970). Such enzymes are monomeric and only require the presence of Mg^{2+} to cleave DNA (Table 2.1). They are defined as *specific endonucleases that recognize particular sites on double-stranded DNA and cut the DNA at (or in the direct vicinity) of these sites.* Type II restriction enzymes reproducibly generate DNA fragments displaying extremities of known nucleotide sequence. For this reason, they are very useful in genetic engineering.

A very large number of type II restriction enzymes that recognize and cleave different DNA sequences have been isolated from several bacterial species. Appendix 2 gives the characteristics of about 475 of them. Their number increases monthly. It is important to bear in mind that it has not been formally proven that most of these enzymes are involved in a host restriction–modification system. In fact, many are specific endonucleases that are inactive on host DNA but are able to hydrolyze exogenous DNA molecules.

A standard nomenclature for restriction endonucleases was proposed by Smith and Nathans in 1973:

- The first capital letter corresponds to the first letter of the bacterial genus from which the enzyme was extracted. The second and third lowercase letters are the first two letters of the species name. (All three letters are in italics.) For instance, *Escherichia coli* is represented by *Eco, Haemophilus influenzae* by *Hin,* and *Haemophilus aegyptius* by *Hae.*

Table 2.1 Properties Of The Different Classes Of Restriction Endonucleases (Adapted From Different Authors)

	Type		
	I	II	III
Restriction and modification	Multifunctional enzyme	Separate endonuclease and methylase activities	Separate enzymes with a common subunit
Structure	3 different subunits	Single protein	2 different subunits
Requirements	ATP, Mg^{2+}, S-adenosylmethionine	Mg^{2+}	ATP, Mg^{2+}, S-adenosyl-methionine
Cleavage site	Random	At or near the recognition site	About 25 bp from the recognition site
Enzymatic turnover	No	Yes	No

- If present, the next letter is not italicized and refers to the strain, bacterial type, virus, or plasmid from which the enzyme was isolated. For instance, *Hin*d is *Haemophilus influenzae* Rd, *Eco*R is *Escherichia coli* RY13, and so on.
- If a given strain contains different restriction–modification systems, they are identified in roman numerals following the published order of isolation (e.g., *Hin*dI, *Hin*dII, *Hin*dIII).

2.3. Properties of the Recognition Site

Table 2.2 shows the *recognition sites* of five type II restriction enzymes. Most known enzymes recognize a sequence of 4 (tetrameric), 6 (hexameric), and sometimes 5 (hexameric) nucleotides. In general, these sites display a rotational symmetry of type 2 and are known as *palindromic sequences.* In other words, the nucleotide sequence of a given strand is identical to that of the complementary strand when they are both read in the 5′ → 3′ direction.

For instance, in the case of *Eco*RI, the recognition sequence is GAATTC in the 5′ → 3′ orientation and the axis of rotation is located between the central AT base pair.

$$5' - \text{GAA} \mid \text{TTC} - 3'$$
$$3' - \text{CTT} \mid \text{AAG} - 5'$$

Cleavage of a DNA fragment at the site recognized by a restriction enzyme can be performed in three different ways. Enzymes such as *Hae*III give rise to *blunt ends* by cutting straight in the middle of the recognition sequence. In contrast, *staggered* (or sticky or cohesive) *ends* are obtained when enzymes yield DNA fragments displaying a protruding strand. This can be either a 5′ phosphate extension (as in the case of *Eco*RI) or a 3′ hydroxy extension (as obtained with *Pst*I).

Table 2.2 Five Commonly Used Restriction Enzymes[a]

Species	Restriction Enzyme	Restriction Site
Escherichia coli RY13	*Eco*RI	G↓AATTC
Haemophilus influenzae Rd	*Hin*dIII	A↓A̅G̅CTT
Haemophilus aegyptius	*Hae*III	G̅G̅↓CC
Klebsiella pneumoniae	*Kpn*I	GGTA̅C̅↓C
Providentia stuartii	*Pst*I	CTGCA↓G

[a] The recognition sites are indicated on a single DNA strand in the 5′→3′ orientation. Cleavage sites are indicated by arrows and methylated bases by underlined letters.

*Hae*III   ↓

5′ . . .GG - 3′   5′ - CC . . . 3′

3′ . . .CC - 5′   3′ - GG . . . 5′

↑

*Eco*RI   ↓

5′ . . .G - 3′   5′ - AATTC . . .3′

3′ . . .CTTAA - 5′   3′ - G . . .5′

↑

*Pst*I   ↓

5′ . . .CTGCA - 3′   5′ - G . . .3′

3′ . . .G - 5′   3′ - ACGTC . . .5′

↑

Isoschizomers are enzymes purified from different microorganisms that recognize the same DNA sequences. This phenomenon is commonly observed among the known restriction enzymes (see Appendix 3.)

EXPERIMENTS

2.1. Purity

Commercially available enzymes are sold in a concentrated form devoid of exonuclease, endonuclease, and phosphatase activities.

2.2. Storage

Restriction enzymes are routinely stored frozen at −20°C in 50% glycerol buffers. On the day of use they should be kept on ice and returned to the freezer immediately after use. Care must be taken not to store restriction enzymes in frost-free freezers, which undergo cycles of heating to remove the ice buildup.

2.3. Unit Definition

A unit corresponds to the amount of restriction enzyme required to digest 1 μg of bacteriophage λ DNA in 1 h at the specified temperature and buffer conditions.

2.4. Measuring Restriction Enzyme Activity

The λ digestion test is performed by DNA electrophoresis on agarose gels. (See chapter 4.) For example, *Eco*RI digestion of λ DNA yields six fragments of

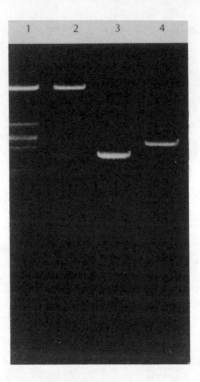

Figure 2.2. *Visualization of different DNA fragments resolved on an agarose gel after staining with ethidium bromide.* The bacteriophage λ DNA contains five *Eco*RI sites located at positions 21,226, 26,104, 31,747, 39,168, and 44,972. Complete digestion with *Eco*RI yields six DNA fragments with molecular weights of 21.8, 7.5, 5.8, 5.5, 4.8, and 3.4 kbp (lane 1). Because of resolution limitations, the 5.8- and 5.5-kbp fragments appear as a single, more intense band. Lane 2 shows adenovirus 2 DNA (35 kbp) digested with *Eco*RI. Digestion also yields six fragments. There is only one *Eco*RI site and thus one fragment on the 4,363-bp-long plasmid pBR322 (lane 3), and one on the 5,243-bp-long SV40 viral DNA (lane 4). There is no *Eco*RI restriction site on the bacteriophage φX174 DNA (5,386 bp).

characteristic sizes when digestion is complete (Fig. 2.2). One must bear in mind that the efficiency of digestion depends on the conformation of the DNA to be cleaved. In the case of plasmid pBR322 in a supercoiled conformation, 5–10 *Eco*RI units may be necessary to achieve complete digestion.

2.5. Reaction Conditions

Enzymatic digestions are generally carried out at 37°C at the exception of a few enzymes that function better at higher temperatures (e.g., *Bst*EII at 60°C). The reaction mixture usually consists of a Tris–HCl buffer in the physiological pH range supplemented with salts such as $MgCl_2$, KCl, and NaCl. The ionic strength, and to a lesser extent the pH, are variable parameters which depend upon the enzyme used. Some restriction endonucleases (e.g., *Hpa*I) work better at low salt concentrations (0–50 mM NaCl), while others (e.g., *Sal*I) require high salt concentrations (150 mM NaCl) for optimum activity. A DNA molecule can be simultaneously digested with two or more different enzymes provided that their buffer compositions do not radically differ. Often, one enzyme will cut suboptimally and it may be necessary to use it at higher concentration or to increase the incubation time. DNA digestions are usually stopped by heating at 65–70°C for 5 min, which is sufficient to inactivate most enzymes, or by addition of a metal ion chelator such as EDTA. Because some enzymes are refractile to temperature inactivation, they must be separated from the cleaved DNA fragments by phenol/chloroform extraction.

Star activity is observed with certain enzymes when the digestion conditions have been modified (e.g., very high enzymatic concentration, Mn^{2+} ions used in place of Mg^{2+} ions; high pH, low ionic strength, high concentration of organic solvents such as glycerol). As a result of star activity, the recognition sequence of the enzyme is modified. For example, the star activity of *Eco*RI (*Eco*RI*) will result in the recognition and cleavage of AATT sequences instead of GAATTC for the normal form of the enzyme.

PROBLEMS

2.1 Enzymes and Extensions

Write the different DNA extremities obtained by cutting with *Hha*I (GCG↓C), *Hpa*II (C↓CGG), *Asu*I (G↓GNCC), and *Eco*RII (↓CC^A_TGG). What is the particularity of the last three enzymes?

2.2 Enzymes With Multiple Recognition Sites

How many and which sequences are recognized by the following enzymes: *Eco*RII (CC^A_TGG), *Scr*FI (CCNGG), and *Afl*III (ACPuPyGT)?

2.3 Rotational Symmetry of Restriction Enzymes

Do you believe that restriction enzymes *Eco*RII (↓CC^A_TGG), *Hinc*II (GTPy↓ PuAC), and *Hae*II (PuGCGC↓Py) follow the second-order rotational symmetry rule? In the affirmative, what is the particularity of these enzymes?

2.4 Isoschizomery

What comments can you make on enzymes *Xho*I (C↓TCGAG), *Sma*I (CCC↓ GGG), *Blu*I (C↓TCGAG), and *Xma*I (C↓CCGGG)?

2.5 Theoretical Frequency of Restriction Cuts

At what theoretical frequency do tetra- or hexameric recognition sequences arise on DNA molecules?

2.6 Theoretical Symmetric Sites

How many symmetrical sites can theoretically arise for restriction enzymes recognizing tetrameric or hexameric DNA sequences?

2.7 Theoretical vs. Actual Recognition Sequences

The left-hand side column of Table 2.3 shows all the possible tetrameric recognition sequences. All possible hexameric recognition sequences can be obtained by combining tetrameric sequences with the two flanking bases shown in the first line of the table. The number of restriction enzymes discov-

Table 2.3 Frequency Of Occurence Of Restriction Enzymes Recognizing Tetrameric Or Hexameric Sequences

Tetrameric		Hexameric			
5′→3′	....	A....T	T....A	C....G	G....C
AATT	0	0	0	0	2
ATAT	0	0	0	0	0
ACGT	0	0	0	0	2
AGCT	2	8	0	1	3
TATA	0	0	0	0	2
TTAA	0	0	0	0	5
TCGA	1	1	0	14	7
TGCA	0	1	0	0	1
CATG	0	0	0	0	0
CTAG	0	0	1	1	0
CCGG	5	0	0	5	0
CGCG	5	0	0	3	0
GATC	8	2	4	4	6
CTAC	1	1	0	0	1
GCGC	5	3	2	0	5
GGCC	12	3	2	1	0

ered with such sequences are indicated at the intersection. What conclusions can you draw from inspection of Table 2.3?

2.8 Enzymatic Units Necessary to Digest a Given Amount of DNA

Derive the formula that relates the number of restriction enzyme units (a) necessary to digest to completion X μg of a plasmid P in 1 h.

CHAPTER

3

Methylation

3.1. Prokaryotic Methylases

K-12, the most widely used *E. coli* strain, contains, in addition to the host restriction–modification methylase, at least two additional methylase activities:

- *dam* methylase (DNA adenine methylase). This methylase introduces a methyl group in position N^6 of the adenine in the GATC sequence (Marinus and Morris, 1973). Nevertheless, no *E. coli* restriction enzyme recognizing this sequence could be isolated. In fact, it is now known that the *dam* methylase plays a role in the error-correction system used by the cell during DNA replication. Briefly, the mismatch repair system of *E. coli* is able to recognize the degree of methylation of a DNA strand carried out by the *dam* gene product and preferentially excises nucleotides from the undermethylated strand. Since the daughter strand is always undermethylated, this system allows *E. coli* to correct the mutations that arise during the replication of the parental strand.

  As a direct consequence *K-12* DNA cannot be digested by the restriction enzyme *Mbo*I (GATC) since its recognition sequence is modified by the *dam* methylase. This applies to both genomic and plasmid DNA (e.g., pBR322) amplified in *E. coli K-12* by the chloramphenicol method.

  Restriction enzyme *Sau*3AI (GATC) recognizes the same sequence as *Mbo*I but it is not sensitive to the action of the *dam* methylase. Hence the GATC sequence can be cut by *Sau*3AI. Similarly, bacterial DNA cannot be digested by *Bcl*I (TGATCA). However, *Bam*HI (GGATCC), *Pvu*I

(CGATCG), *Bgl*II (AGATCT), and *Xho*I (PuGATCPy) can cut even if the internal adenine is methylated. If an *E. coli* mutant deficient in *dam* methylase activity is used to isolate a DNA of interest (whether genomic or plasmidic), *Bcl*I (TGATCA) digestion then becomes possible.

Finally, certain restriction-enzyme recognition sequences can be protected from cleavage by the *dam* system. A typical example is the enzyme *Taq*I (TCGA). If this sequence is followed by the TC nucleotide pair, the adenine of the TCGATC sequence will be methylated by *dam* and *Taq*I will be unable to cut at this site.

- *dcm* methylase (<u>D</u>NA <u>c</u>ytosine <u>m</u>ethylase). This methylase introduces a methyl group at the C^5 position of the second cytosine in the CCAGG or CCTGG sequence (May and Hattman, 1975). The enzyme *Eco*RII recognizes either of these sequences and is sensitive to *dcm* methylation. In contrast, *Bst*NI, that shares the same recognition sequence, is insensitive to cytosine methylation and still cuts. Some *dcm* *E. coli* mutants are also available to circumvent this problem.

3.2. Methylation by Eukaryotes

Eukaryotic DNA is very resistant to digestion by restriction enzymes that contain the 5′-CG-3′ nucleotide pair in their recognition sequence [e.g., *Hpa*II (CCGG)]. This is very likely due to the following particularities:

- In mammals the cytosine of the CG doublet is often methylated. This protects DNA sequences containing such nucleotide pairs from cleavage by restriction enzymes containing CG in their recognition sequence.
- Cytosines are subject to hypermutations and the CG doublet is rarer than would be expected. Appendix 4 gives a list of the sequences that are or are not digested by different enzymes depending upon the methylation of adenines or cytosines in the recognition sites. Note that *Hpa*II cuts C<u>C</u>GG (underlined letters denote methylated nucleotides) but not C<u>C</u>GḠ. The opposite is true for the enzyme *Msp*I. Thus, this pair of enzymes can be used to determine the number and location of methylated cytosines in different DNA samples (Waalwijk and Flavell, 1978).

EXPERIMENTS

3.1. Example: *Eco*RI Methylase

E. coli *Eco*RI methylase catalyzes the transfer of a methyl group from *S*-adenosylmethionine to the underlined adenine in the sequence GA<u>A</u>TTC. This sequence is recognized by the restriction enzyme *Eco*RI but is not cleaved if the adenine has been methylated (Greene et al., 1975).

3.2. Unit Definition

A unit of *Eco*RI methylase is the quantity of enzyme necessary to protect 1 μg of λ DNA from *Eco*RI digestion for 1 h at 37°C in the appropriate buffer.

3.3. Reaction Conditions

The DNA to be methylated is mixed with *S*-adenosylmethionine in 100 mM Tris-HCl, pH 8.0, containing 10 mM EDTA.

PROBLEMS

3.1 Methylation Sites

Refer to Appendix 4 for the sequences methylated by the *Taq*I and *Bam*HI methylases. Underline the methylated bases.

3.2 Selecting Restriction Enzymes to Study Adenine Methylation

Refer to Appendix 4 to find three restriction enzymes that can be useful in the study of adenine methylation.

3.3 Effect of the *dam* Methylase on Restriction Enzymes

The restriction enzymes *Cla*I (AT↓CGAT), *Xba*I (T↓CTAGA), and *Hph*I (GGTGA) are sensitive to *dam* methylation. Write the complete sensitive sequences for each of these enzymes.

3.4 Protecting Restriction Sites by Methylation

Show how the *Msp*I methylase protects specific *Bam*HI sites in the sequence 5′ -TGCCGGATCCTGATGGATCCT-3′.

3.5 Methylation and Cleavage of Eukaryotic DNA

Explain why eukaryotic DNA is resistant to *Sma*I digestion.

4

Electrophoresis and Blotting Techniques

Most biological molecules are charged species that can move in an electrical field. Electrophoresis techniques are based upon this property. This chapter describes some of the most important applications of this method for the separation and analysis of DNA and proteins.

4.1. DNA Electrophoresis

Under neutral or alkaline pH conditions, the phosphate groups of DNA chains become ionized into *polyanions.* As a consequence, DNA fragments placed within a porous matrix and subjected to a difference in electrical potential travel toward the anode. Under these conditions, small linear fragments or tightly packed, *supercoiled* circular DNA molecules migrate faster through the pores of the gel relative to their larger or more loosely packed counterparts. In this respect, the porous matrix acts as a molecular sieve. By using this technique, known as *gel electrophoresis,* it becomes possible to separate DNA molecules of different size or conformation. The main factor affecting the mobility of a DNA fragment is the viscosity of the matrix in which it is placed.

Two types of gel matrices are used for DNA electrophoresis—*agarose* and *acrylamide* (Loening, 1967; Aaij and Borst, 1972). When melted in a suitable buffer and allowed to solidify, agarose, a material derived from seaweeds, forms a porous, gel-like matrix. The size of the pores depends on the agarose-to-buffer ratio, which can be varied from 0.5 to 1.5%. Another type of matrix, *poly-acrylamide,* can be obtained by forming a copolymer of acrylamide and

Table 4.1   Influence Of The Agarose Concentration On The Resolution Of Linear
               DNA Fragments

% Agarose in the Gel	Optimal Separation Range of Linear DNA Fragments
0.6	20–1.0 kbp
0.7	10–0.8 kbp
0.9	7–0.5 kbp
1.2	6–0.4 kbp
1.5	4–0.2 kbp

bis-acrylamide. The polymerization reaction is initiated by addition of
N,N,N',N'-tetramethylethylenediamine (TEMED) and catalyzed by ammonium persulfate. As in the case of agarose, the size of the pores in the gel matrix
can be modified widely by varying the acrylamide content and the acrylamide–
to–bis-acrylamide ratio. In general, agarose gels display larger pores than polyacrylamide gels. Hence, polyacrylamide gels are essentially used to separate
DNA fragments of molecular weight inferior to 1 kbp. In contrast, agarose gels
can resolve DNA fragments ranging between 0.5 and 50 kbp (Table 4.1). For
example, a 12-cm 1% agarose gel will separate fragments between 30 kbp and
200 bp, while a 12-cm 7.5% polyacrylamide gel will only resolve DNA molecules ranging between 50 bp and 2 kbp in length.

  Ethidium bromide (Fig 4.1) is a base analog that can reversibly intercalate
between the bases of DNA molecules. It emits an orange fluorescence upon
excitation by UV light. DNA fragments resolved by electrophoresis can be
visualized by incubating the gel in a solution containing ethidium bromide and
exposing it to UV light (Sharp et al., 1973; Fig. 4.2). Ethidium bromide fluo-
resces optimally at 254 nm. However, DNA molecules can be cut and dimer-

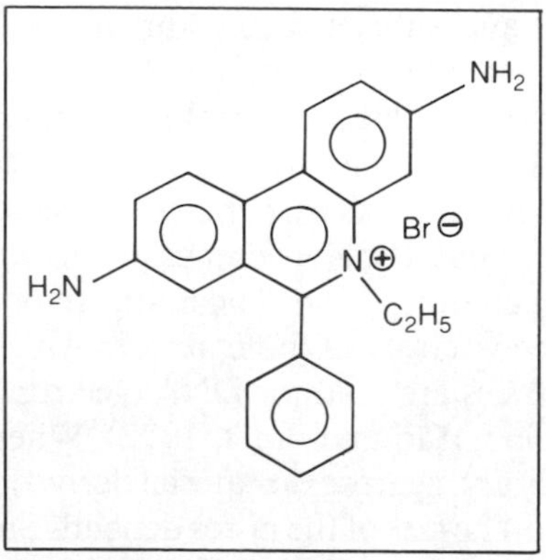

Figure 4.1.   *Chemical formula of ethidium bromide.*

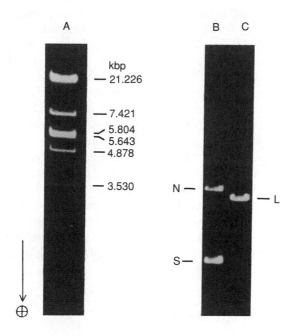

Figure 4.2. *Effect of DNA size and conformation on electrophoretic mobility.* Different samples were resolved on 1% agarose gels, stained with ethidium bromide, and photographed on a UV transilluminator. Lane A: bacteriophage λ DNA digested with *Eco*RI. The exact size of the fragments, as determined by sequencing, are shown. Lane B: nicked (N) and supercoiled (S) forms of a 6.4-kbp plasmid. Lane C: the plasmid of lane B following linearization with *Eco*RI (one side). Note that the linearized (L) form migrates slightly faster than the nicked (N) form under the electrophoresis conditions used for this experiment.

ized at this wavelength (Brunk and Simpson, 1977). Thus, polyacrylamide or agarose gels are generally visualized at 300 nm.

In addition to the nature and viscosity of the gel matrix used, the electrophoretic mobility of DNA fragments depends on two factors (Fig. 4.2):

- *Size.* As already mentioned, double-stranded linear DNA molecules migrate faster through a gel as their size decreases. In fact there is a rough logarithmic correlation between the molecular mass (M_r) of a DNA fragment and the distance it will travel in a gel.
- *Conformation.* Plasmid DNA is isolated in a circular form that can adopt different conformations. The *supercoiled* form consists of tightly packed, highly twisted circular molecules. The *covalently closed* form is characterized by a relaxed circular conformation that can lie flat on a surface. If some phosphodiester bonds are broken in one or the other DNA strand of the previous

form, the conformation is referred to as *nicked*. In addition, *linearized* plasmid DNA may be obtained by cleavage with restriction enzymes. In most cases, a supercoiled DNA molecule will migrates faster on an agarose gel than its linearized counterpart. However, a number of other factors can affect the migration pattern of a given DNA molecule. These factors include the diameter of the gel pores, the voltage used, the ionic strength of the buffer, and the dye concentration. For instance, nicked DNA incorporates larger amounts of ethidium bromide when this dye is used at high concentration. This phenomenon results in a progrcssivc transformation of thc nicked form into the supercoiled form, and thus an increased mobility in the gel. The analysis of plasmid preparations is further complicated by the existence of supercoiled or relaxed dimeric forms. Hence, the identity of a plasmid is generally confirmed by digesting it with a restriction enzyme recognizing a unique site within the plasmid, and by comparing the migration pattern of the linearized molecule to that of standards of known molecular weights. Typical molecular weight standards are bacteriophage λ DNA digested with *Hin*dIII or bacteriophage ϕX174 DNA digested with *Hae*III.

4.2. Protein Electrophoresis

The separation of proteins by sodium dodecylsulfate–polyacrylamide gel electrophoresis *(SDS-PAGE)* has had a significant impact in genetic engineering. The technique was originally developed as a means to measure the molecular weight of proteins. When a protein sample, dissolved at neutral pH in a buffer containing 0.1 M mercaptoethanol and 1% SDS, is heated at 100°C for a few minutes, the subunits of globular proteins dissociate, the disulfide bonds are broken by the reducing agent, and all secondary structure is lost. The negatively charged detergent SDS binds to the resulting random coils at a constant amount per unit weight of protein (1.4 g of SDS per g of protein). As a result, the proteins behave as if they had a uniform shape and identical charge-to-mass ratio. Although one would expect that the mobility of such modified proteins would be directly proportional to their molecular weight, this is not absolutely true. As already mentioned, gel matrices such as polyacrylamide behave as a molecular sieve in which small molecules can travel faster than larger ones. As a result, there is no direct linear relationship between distance migrated and molecular weight. However, when a mixture of proteins of known molecular weight is run in parallel with the protein of interest, the molecular weight of the latter may be estimated from a semilogarithmic plot of the molecular weight of the standard set of proteins vs. the distance they have migrated in the gel. This technique yields molecular weights with a 5–10% accuracy.

SDS–polyacrylamide gels consist of a *stacking* domain followed by a *resolving* domain (Fig. 4.3). The stacking portion of the gel (pH 6.7), in which wells are generated using a comb, has a very large pore size and thus a low impedance. (The acrylamide concentration is usually 3–5%.) Protein separation is

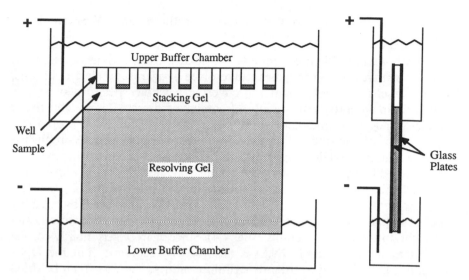

Figure 4.3. *Schematic representation of a SDS-PAGE setup.*

achieved in the resolving portion (pH 8.9). Typically, the acrylamide concentration varies between 8 and 15% depending on the molecular weight of the proteins for which the best resolution is sought. Protein samples (usually 10–40 μg) are loaded in the wells of the stacker, the whole assembly is immersed in a Tris-glycine buffer at pH 8.3, and an electric field is applied. As the proteins start migrating in the well, they become "sandwiched" between chloride and glycine ions, and pack as a thin, stacked layer at the beginning of the resolving domain of the gel. This technique minimizes the spreading of the samples and yields thin and well-resolved protein bands. The progression of electrophoresis can be visually followed by adding bromophenol blue to the protein sample. Upon completion of electrophoresis, the proteins are stained using a Coomassie blue dye that highlights bands containing about 1 μg of proteins. For the detection of bands containing very little protein (1 ng) a silver-nitrate-based stain is used. Once resolved by electrophoresis, proteins may be extracted from gel slices for other experiments (e.g., to obtain antibodies against a particular protein). Protein electrophoresis is often a first step for many applications. For instance, if the proteins separated by electrophoresis have been radioactively labeled, the gel can be dried and layered on top of an X-ray film in order to generate an *autoradiogram*. Another very important application, immunoblotting, is described in the following section.

4.3. Blotting Techniques

The first transfer procedure was developed for DNA by Edwin Southern and is known as Southern blotting. Similar techniques were soon developed for RNA and proteins and humorously named Northern and Western blots, respectively.

- *Southern blotting.* When genomic DNA isolated from various organisms is digested with restriction enzymes and resolved by agarose gel electrophoresis, a long smear of DNA fragments covering a whole spectrum of size is obtained. In order to identify a DNA fragment of interest, it is possible to use single-stranded and radiolabeled RNA or DNA *probes.* Such probes can readily *hybridize* to single-stranded DNA, but this is not easily accomplished in a gel. The method developed by Southern consists of denaturing the electrophoresed DNA by incubating the gel in a NaOH solution and transferring the DNA fragments onto a sheet of nitrocellulose while maintaining the electrophoretic resolution pattern. This is practically accomplished by covering the gel with a sheet of nitrocellulose cut to fit and stacking paper towels and a weight on top of the assembly (Fig. 4.4). Since agarose gels mainly consist of water, the liquid (and DNA) is absorbed by capillarity but the DNA remains tightly bound to the nitrocellulose matrix. The latter can be dried and a specific probe can be used to detect the DNA fragment of interest by autoradiography of the nitrocellulose filter.
- *Northern blotting.* Under specific conditions, RNA can be transferred to nitrocellulose filters by using the Southern blot setup. The RNA is denatured by treatment with glyoxal and DMSO and electrophoresed on an agarose gel by using a recirculating buffer. Transfer to nitrocellulose is carried out in a

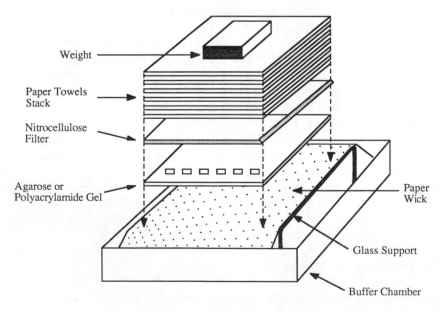

Figure 4.4. *Schematic representation of a Southern blot setup.*

high salt buffer as described above. The dried filters can be used directly for hybridization.

- *Western blotting (or immunoblotting).* The transfer of proteins resolved by SDS-PAGE to nitrocellulose paper is not as straightforward as in the case of DNA and requires additional equipment. Briefly, the SDS gel is washed in transfer buffer, placed on top of a filter paper, and covered with a sheet of nitrocellulose cut to fit. Another filter paper is placed on top of the nitrocellulose and the whole assembly is placed between two foam pads and inserted in the holder of an electrophoretic cell (Fig. 4.5). It is important that the nitrocellulose filter face the anode since when an electrical field is applied, the negatively charged SDS–protein complexes will migrate in this direction. Since the blotting electrical field is perpendicular to that used for SDS-PAGE, the proteins will maintain their pattern as they exit the gel and bind to the nitrocellulose. Once transfer is complete, the nitrocellulose filter is washed and blocked by incubation in a gelatin or dry milk buffer. The protein of interest is then immunologically detected by incubating the membrane with a buffer containing the appropriate *primary antibody.* Following a wash step, the filter is incubated in buffer containing a *secondary antibody* raised against the primary antibody and conjugated to a chromogenic enzyme (e.g., a goat anti-

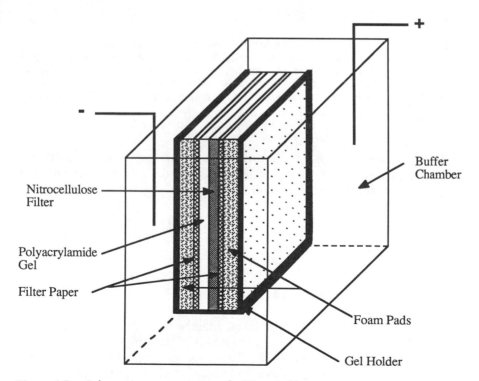

Figure 4.5. *Schematic representation of a Western blot setup.*

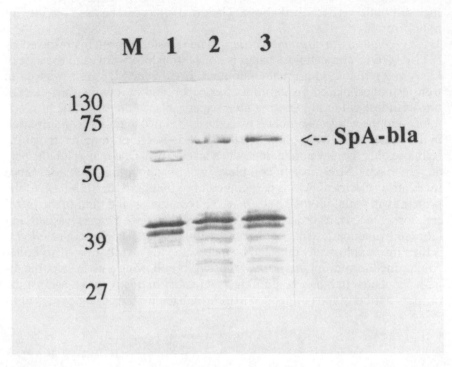

Figure 4.6. *Typical Western blot.* Different *E. coli* strains (lanes 1–3) harboring a plasmid coding for a protein A-β-lactamase gene fusion were grown to midexponential phase; an identical amount of periplasmic cell extract was resolved by electrophoresis and detected with an anti-β-lactamase antibody. The position of the fusion (SpA-bla) is indicated by an arrow. Lane M is a set of prestained molecular weight markers (in kDa).

IgG-peroxidase conjugate if the primary antibody was raised in a rabbit). When the chromogenic enzyme substrate is added, bands corresponding to the protein of interest will appear (Fig. 4.6).

EXPERIMENTS

4.1. Electrophoresis Cells

• *Vertical devices.* Such cells are used mainly with polyacrylamide gels, although agarose matrices may also be employed. The migration of DNA and proteins is fast and reproducible. This type of apparatus is principally employed for the resolution of low-molecular-weight DNA fragments, DNA

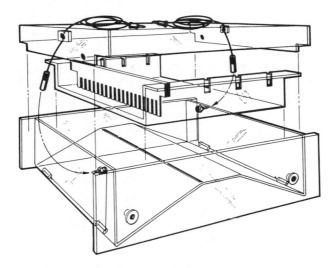

Figure 4.7. *Example of a horizontal electrophoresis cell* (model H4 from BRL).

sequencing, and virtually all SDS–polyacrylamide gel electrophoresis of proteins.

· *Horizontal devices.* Horizontal cells are used exclusively with agarose matrices. The system is easy to set up and operate and allows the use of low-percentage agarose gel in order to resolve large-molecular-weight DNA fragments (Fig. 4.7). At present, the most popular horizontal electrophoresis

Figure 4.8. *Example of a minigel system* (model H6 from BRL).

apparatus is the minigel system (Kopchik et al., 1981; Fig. 4.8). Small agarose gels are poured into UV-transparent trays, placed in the electrophoresis tank, and immersed under 1–2 mm of buffer containing ethidium bromide. The migration of the DNA fragments dyed by ethidium bromide can be easily followed in the dark by illuminating the gel with a hand-held UV lamp. Minigels present a number of advantages. Since the gels are small, little agarose is used. Furthermore, as a consequence of the high conductivity of the "submarine" setup, results are obtained rapidly. (*Eco*RI-digested λ DNA fragments are resolved in 20 min with a minigel compared to 2 h with a vertical system.) Minigels have proven helpful for many tests requiring rapid answers (e.g., presence of DNA, verification of digestion completion, etc.). To increase the resolution of DNA fragments of similar size, a larger version of the minigel system can be used.

4.2. Agarose

Type II agarose is the common matrix for DNA electrophoresis. The gels obtained are transparent and mechanically stable at low agarose concentration, and the presence of sulfur polysaccharides in the agarose inhibits the action of contaminating restriction enzymes, ligases, and polymerases during electrophoresis.

4.3. Electrophoresis Buffers

Three different types of buffer solutions may be employed for DNA electrophoresis:

- *Tris-acetate buffer* (TAE). The 10× stock solution consists of 400 mM Tris, pH 7.8; 200 mM sodium-acetate; and 18 mM EDTA. TAE requires a buffer circulation between the tanks at the anode and the cathode in order to equilibrate the pH.
- *Tris-borate buffer* (TBE). The 10× stock is 890 mM Tris base, pH 8.3; 890 mM boric acid; and 25 mM EDTA. TBE is the most widely used electrophoresis buffer although it may damage the gel through the formation of borate–agarose complexes. Furthermore, DNA is difficult to elute out of this type of gels.
- *Phosphate buffer.* The buffer consists of either 50 mM Tris/NaH_2PO_4, pH 7.5–8.0, or of 50 mM Na_2HPO_4/NaH_2PO_4, pH 7.5–8.0.

  For protein electrophoresis a 4× stock solution of running buffer is prepared by dissolving 12 g of Tris/base and 57.6 g of glycine base in 1 L of ddH_2O.

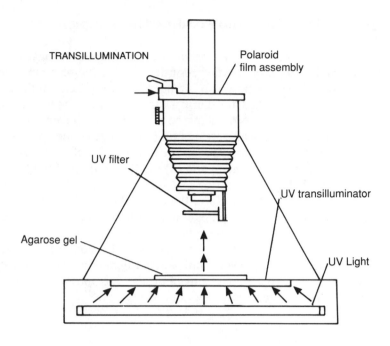

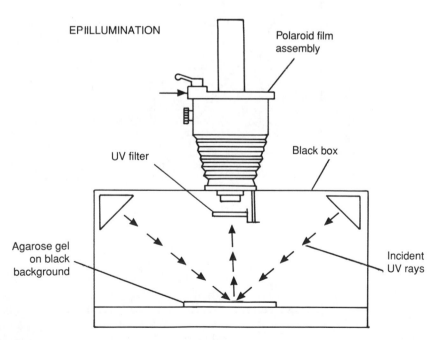

Figure 4.9. *Epi- and transillumination of stained agarose gels* (from Maniatis et al., 1982).

4.4. Casting DNA Gels and Loading the Samples

Agarose gels are prepared by melting the desired amount of agarose in the selected electrophoresis buffer. After cooling to 50°C, the solution is poured into a special tray and a plastic comb is placed toward one end of the gel (Fig. 4.7). Once the gel has solidified, the comb is carefully removed and the resulting wells can be filled with the DNA sample to be analyzed. The amount of DNA that can be loaded in each well depends on (1) the size of the comb tooth (it must, however, remain inferior to 50 μg/cm^2), (2) the size of the DNA (loading capacity decreases with DNA size), and (3) the distribution of the DNA fragments. In general, 5–15 μg of digested DNA is loaded per well in large agarose gel. The loading buffer used to dissolve the DNA of interest consists of a denaturing agent (SDS or urea) to stop possible enzymatic reactions, a dense compound (glycerol, Ficoll, or sucrose) to guarantee that the sample remains in the well, and mobility markers (bromophenol blue, which comigrates with small restriction fragments, and xylene cyanol, which migrates with large fragments) in order to follow the progress of electrophoresis. The optimal voltage for the resolution of DNA fragments is 5 V per cm of gel.

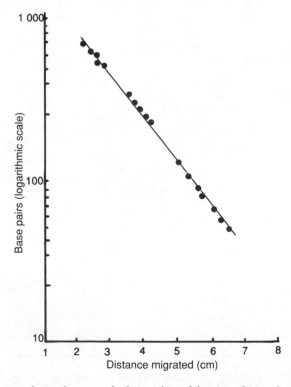

Figure 4.10. *Correlation between the logarithm of the size of DNA fragments obtained by Hae III digestion of pBR322 and the electrophoretic mobility on an acrylamide gel.*

4.5. Visualization of DNA Bands

As mentioned previously, DNA fragments stained with ethidium bromide can be observed upon exposure to UV light at 300 nm. The maximum emission wavelength is 590 nm. Once the gel is stained, visualization and photography can be performed either by *epiillumination* or *transillumination* (Fig. 4.9).

4.7. Migration Anomalies

- *Edge effects.* They result in "smiley" gels when the voltage applied is too high for the gel concentration used.
- *Overloading effect.* If too much DNA is loaded per well, migration of the restriction fragments will be slowed down.
- *Contamination.* The migration of restriction fragments is similarly retarded when samples are contaminated by salts or proteins. Such impurities can be

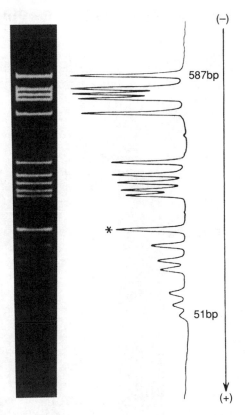

Figure 4.11. *Densitometric scanning of the DNA fragments obtained by digesting pBR322 with HaeIII and subjecting the mixture to electrophoresis on an acrylamide gel.*

removed by phenol extraction (proteins) or successive ethanol precipitations (salts) of the sample.

PROBLEMS

4.1 DNA Molecular Weight Markers for Agarose Gels

When bacteriophage λ DNA is digested with the enzyme *Hind*III, the following restriction fragments are obtained: 23.1, 9.4, 6.5, 3.5, 2.3, 2.0, and 0.5 kbp. Upon electrophoresis on a 1% agarose gel, staining with ethidium bromide, and exposure to UV light, the distances traveled by these fragments are found to be 3.5, 4.5, 5.0, 6.0, 8.0, 8.5, and 12.5 cm, respectively. Draw a plot of fragment size vs. distance migrated. What happens to the curve if the base-10 logarithm of the size is plotted vs. the distance traveled? What are your conclusions?

4.2 DNA Molecular Weight Markers for Polyacrylamide Gels

Figure 4.10 shows the relationship between the logarithm of the size of *Hae*III-digested pBR322 DNA and the mobility of the fragments on a 7.5% polyacryl-

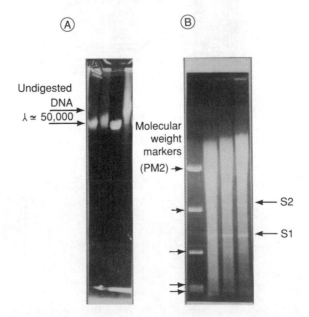

Figure 4.12. *Comparison of the migration patterns of long-strand rat liver DNA undigested (panel A) or digested with EcoRI (panel B).*

amide gel. What conclusions do you derive from this plot? Figure 4.11 shows the position of the fragments on the gel following ethidium bromide treatment as well as the corresponding densitometric recording. What can you say of the band marked with an asterisk?

4.3 Criteria for Complete Digestion of Eukaryotic DNA

How many restriction fragments are obtained upon digestion of mammalian DNA (10^9 bp) by *Eco*RI? Figure 4.12B shows the migration pattern obtained when 10 μg of rat DNA digested with *Eco*RI is resolved on an agarose gel. (The first lane of the gel shows molecular weight markers from phage PM2.) What can you say about the pattern? How do you explain the discrete bands labeled S1 and S2?

Part III
Cloning Vectors

CHAPTER
5

Bacteriophage λ

Bacteriophage λ, or phage λ (or even λ), is an *E. coli* phage. Its molecular genetics, particularly replication, gene regulation, and morphogenesis, have been studied extensively. It is hence not surprising that it was one of the first vectors used for cloning purposes. Bacteriophage λ is a double-stranded DNA virus encoding about 50 genes whose 48,502-bp sequence has been determined (Sanger et al., 1982). It encodes, at each of its ends, two short complementary DNA sequences of 12 nucleotides (5'-GGGCGGGCGACCT-3') which constitute the *cos cohesive ends*.

5.1. Genes Involved in the Alternative Life Cycles of λ

λ is a *temperate* phage that can undergo two different life cycles: *lytic* and *lysogenic* (Fig. 5.1). Both pathways eventually result in viral replication. The lytic phase consists in the multiplication of phage particles and the subsequent destruction of the host cell. The lysogenic pathway is characterized by the integration of the phage genome within the *E. coli* chromosome. The resulting *prophage* replicates with the *E. coli* genome and remains genetically silent. Upon damage to the host chromosome (e.g., through UV irradiation), λ is excised and follows the lytic pathway.

Upon infection, λ injects its linear DNA in the host cell. Once inside the cytoplasm, the phage DNA circularizes through its cohesive, complementary ends to form the *cos* site. One of the two following pathways is then followed

47

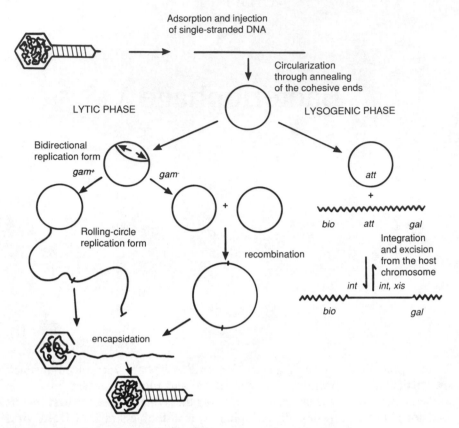

Figure 5.1. *Alternative life cycles of the temperate bacteriophage λ.*

depending upon the interactions of the *cI, cII,* and *cro* gene products with the operators of phage λ.

- *Lytic pathway.* The phage circular DNA molecules replicate bidirectionally. Initially, replication originates between the *cII* and *O* genes and involves genes *O* and *P*. At the time of infection, a *rolling circle* type of replication is adopted. This results in the formation of long *concatemeric* molecules composed of successive genomes joined to each other. The product of the *gam* gene, which inhibits the *E. coli* nuclease *rec*BCD, is involved in the switch between replication modes.

    Mutant *gam⁻* phages are only able to produce concatemeric molecules in *rec*BCD⁻ hosts. Because the formation of such molecules is required for the encapsidation of phage DNA, *gam⁻* phages infecting *rec*BCD⁺ *E. coli* use recombination to produce concatemers. This alternative route involves the phage *red* and the *E. coli recA* gene products.
- *Lysogenic pathway.* Integration of λ within the *E. coli* genome involves a

recombination event at the attachment site *(att)* of the phage. This site is located between the genes coding for galactose utilization *(gal)* and biotin biosynthesis *(bio)* and shares high homology with another site located on the host chromosome. Integration of the phage DNA requires the expression of gene *int*. This process is reversible and excision of the prophage involves the combined action of the *xis* (for excision) and *int* (for integration) gene products.

In the prophage form, the only functional gene is *cI*. This protein represses all the genes involved in the lytic pathway.

5.2. Gene Map

A simplified map of bacteriophage λ is provided in Fig. 5.2. (For a more detailed map, see Appendix 5.) It is interesting to notice that functionally related genes are physically grouped on the genome, with the exception of the regulatory genes *N* and *Q*. The main genes involved in the phage head (*A* to *F*) and tail (*Z* to *J*) assembly are found in the first third of the genome. The genomic region located between the genes *J* and *att* consists of *nonessential* genes that encode unknown or secondary proteins.

Genes located on the right-hand side of *att* are *int* and *xis* (involved in the lysogenic pathway) and *red*A and *red*B (involved in recombination events). None of the genes located between *J* and *N* are essential in the lytic pathway that eventually leads to plaque formation. The *N* gene product is important in the regulation of early transcription while the *cI* gene encodes the structural repressor. The latter polypeptide renders cells that are lysogenic for λ immune to further infection *(superinfection)* by another λ phage. It is also responsible for plaque turbidity.

In summary, all the genes involved in the lytic pathway are only located in the right- and left-hand-side portions of the genome.

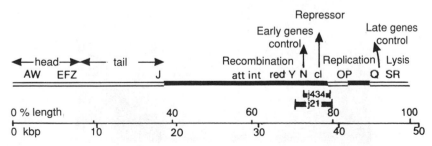

Figure 5.2. *Simplified genetic map of the λ genome.* The solid areas correspond to nonessential genes. The positions of two useful immunity substitutions (*imm*[434] and *imm*[21]) are shown.

5.3. Plaque Formation

When about 10^8 bacteria are grown on a thin section of top agar (about 1 mm thick), a continuous layer of confluent cells known as *bacterial lawn* is obtained. If phage particles are added to the bacteria, a few cells will be infected. When the phage enters the lytic phase and releases its progeny from the host, neighboring bacteria are in turn infected. This process yields a series of clear areas about 1 mm of diameter that appear as clear spots on the turbid bacterial lawn. Such areas are called *plaques.* They consist of lysed bacterial cells and contain on the average from 10^8 to 10^9 phage particles (Fig. 5.3).

Bacteriophage λ adsorbs onto bacterial cells through a specific receptor involved in the maltose fermentation pathway. When bacteria are grown in the presence of maltose, the maltose operon is induced and the amount of the receptor is increased at the surface of the cells. Consequently, growth on maltose maximizes infection by λ. In contrast, growth on glucose must be avoided since it suppresses the formation of the receptors. Finally, magnesium is essential for phage λ stability and it is necessary to include this ion, in the form of magnesium chloride or magnesium sulfate, in all the solutions used for phage storage, dilution, and infection.

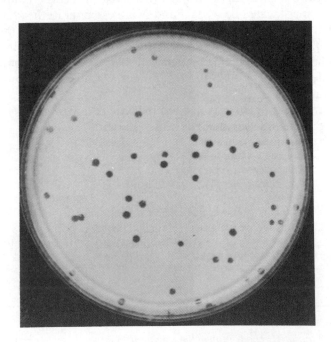

Figure 5.3. *Bacteriophage plaques.* On a dark background, the clear plaques generated on a confluent bacterial layer by bacteriophage λ appear as black spots.

EXPERIMENTS

5.1. Infection and Plaque Formation

Prepare an overnight bacterial culture by inoculating 50 ml of LB supplemented with 0.2% maltose with a single colony. The next day spin down the
cells at 4,000g for 10 min at room temperature and resuspend the pellet in 10
mM MgSO$_4$. The suspension can be kept for about a week at 4°C. Store and
dilute an appropriate bacteriophage stock in SM medium (5.8 g NaCl; 2 g
MgSO$_4$·7H$_2$O; 50 ml Tris-HCl, pH 7.5; 5 ml of 2% gelatin; ddH$_2$O up to 1 L).
On the day of use make 10-fold serial dilutions of the phage stock and mix 100
μl of the dilutions with 100 μl of bacteria at a 10^8 cells/ml concentration. Allow
the phage to absorb by incubating the mixture at 37°C for 15 min. Add 3 ml of
top agar prewarmed to 45°C (10 g Bacto-tryptone, 5 g yeast extract, 2.5 g NaCl,
and 2.64 g MgSO$_4$·7H$_2$O and 0.7% agar per liter of ddH$_2$O). After vortexing,
pour into a fresh petri dish containing 25 ml of *bottom agar* (same composition
as above except for a concentration of agar of 1.2%). Allow the top layer to
solidify at room temperature for 5 min, invert the plates, and incubate at 37°C
overnight. Plaques will start appearing after 8 h incubation and can be counted
and picked after about 16 h.

5.2. Plaque Excision

Using a cut-off Pasteur pipette connected to a plastic bulb, recover the desired
plaque by excising the top and bottom agar surrounding it. Wash the sample
in 1 ml of SM medium containing one drop of chloroform. Let sit for 1 h at
room temperature to allow the phage particles to diffuse from the agar. In general from 10^6 to 10^7 phage particles can be collected by this technique and be
stored at 4°C for a few weeks without loss of viability.

5.3. Preparation of Lysate Stock Solutions

Two methods may be used: liquid storage and Blattner's technique.

For the first method, dilute the desired bacterial culture 1:50 in LB medium
supplemented with 1 mM MgSO$_4$. Grow the cells to OD$_{650}$ = 0.3 at 37°C with
agitation. Infect the culture with a dilution of the phage lysate corresponding
to 2–3 × 10^8 *plaque-forming units* (PFU) per ml. Follow cell growth until a
decrease in OD is observed. When the OD has reached a minimum, add 0.5 ml
of chloroform for every 250 ml of culture and agitate for 15 min. Clarify the
lysate by centrifugation at 10,000 rpm for 10 min. Titrate on appropriate recipient strains by using several dilutions of the stock.

For the second method, mix a plaque removed from the lawn by using a

Pasteur pipette (about 6×10^5 PFU/ml) with 0.3 ml of stationary-phase bacterial culture and 0.3 ml of 10 mM $MgCl_2$, 10 mM $CaCl_2$. Preadsorb the phage to the cells without agitation for 10 min at 37°C. Add 250 ml of LB medium containing 1 mM $MgSO_4$ and grow the cells until a decrease in OD is observed. Proceed as described above.

With both methods, about 10^{10} PFU/ml stock lysates are routinely obtained. The phage stock may be concentrated by centrifugation at 45,000g for 3 h at 4°C and resuspended overnight in 10 ml of SM buffer at 4°C with slow agitation. The next day, clarify the supernatant by centrifugation at 10,000 rpm for 10 min at 4°C.

PROBLEMS

5.1 Stages in the Lytic Cycle and Encapsidation

Using a figure, describe the different steps of the bacteriophage λ lytic pathway. Show the molecules that are likely to be encapsidated.

5.2 Phage Adsorption and Entry

Why are phage particles plated first at room temperature and then transferred to 37°C? What use is this preliminary step with respect to plaque size.

5.3 Increasing Plaque Size

What conditions are likely to increase plaque size?

CHAPTER

6

Bacteriophage Cloning Vectors

Wild-type bacteriophage λ cannot be used directly as a cloning vector since it contains several routinely used restriction-enzyme sites located in genes that are essential for the lytic pathway. In addition, efficient encapsidation can only be achieved with a DNA molecule whose size is comparable to that of the bacteriophage genome itself. Hence, only a small amount of genetic material could be cloned in a wild-type λ phage. For these reasons, a series of genetically engineered bacteriophages, specifically designed for the cloning of extraneous DNA have been developed. The strategies used for the design of these vectors and the description of some important bacteriophage vectors are given in this chapter.

6.1. Removal of Restriction Sites

The central third of the bacteriophage λ genome (genes J to N) is not essential for the lytic pathway, suggesting that it could be removed and substituted by extraneous DNA. Indeed, transduction experiments showed that this domain can be replaced by in vivo genetic manipulation with a wide variety of foreign DNA (e.g., from *E. coli*). This property opened the door for the construction of cloning vectors derived from bacteriophage λ.

Wild-type λ contains five *Eco*RI sites (which we will number 1 to 5 from left to right; see Appendix 6). These sites are located on the left arm of the phage in the nonessential region. Therefore, several λ derivatives containing a smaller number of *Eco*RI sites have been constructed by using mutants carrying deletions at sites 1 and 2 or substitution at site 3 (Murray and Murray, 1974; Ram-

53

bach and Tiollais, 1974; Thomas et al., 1974). Furthermore, by using alternative growth of λ on *E. coli* K carrying or lacking the *Eco*RI restriction–methylation system, it became possible to obtain λ mutants lacking all of these sites. By crossing such restriction-site mutants with *Eco*RI-containing phages, mutant bacteriophages containing site 1 only, site 2 only, or sites 1, 2, and 3 were isolated.

The elimination of multiple restriction sites was carried one step further with the engineering of phages lacking *Hin*dIII sites (Murray et al., 1977). Although wild-type λ has six *Hin*dIII sites, the deletion of the *Eco*RI fragment located between sites 1 and 2 results in the removal of the first two. In addition, phages carrying the *b*538 deletion have lost their first three *Hin*dIII sites, and substitution of the immunity region *(imm)* of λ by that of phage 21 (*imm*[21]) eliminates the *Hin*dIII sites 4 and 5 (using the same convention). Finally, the sixth *Hin*dIII site can be removed by substituting the corresponding genomic region of λ with that of ϕ80 (another lambdoid phage).

6.2. Phage Vectors Derived From λ

The *encapsidation* process consists of the formation of a complex between proteins and DNA in mature phages. Specific nucleases located at the base of the phage head make DNA encapsidation possible by cutting the DNA at the *cos* site. This process is strictly limited with respect to the size of the DNA to be encapsidated. In fact, viable phage particles can only be obtained when the DNA represents 78 to 105% of the size of the wild-type genome (i.e., between 38.3 and 51.6 kbp; Weil et al., 1973). Since 30 kbp of the λ genome consists of essential genes, only those extraneous DNA fragments smaller than 22 kbp can be cloned in λ vectors.

Depending on the specific genome of the λ-derived phages, there is both a lower and an upper limit to the size of extraneous DNA that can be cloned. Two sets of λ-derived vectors have been developed for cloning purposes:

- *Insertion vectors.* Such vectors are used for the cloning of small DNA fragments. The foreign DNA is inserted at a unique restriction site located in a nonessential region of the genome.
- *Substitution vectors.* This class of vectors is used when large DNA fragments (>10 kbp) are to be cloned. A *central substitution fragment* (also known as a stuffer) located between two identical restriction-enzyme sites is removed from the phage and replaced by the DNA of interest. Since there is a lower size limit necessary to achieve DNA encapsidation, the removed fragment must be replaced and cannot simply be deleted.

6.3. Categories of λ Vectors

Appendix 5 shows that wild-type λ carries unique sites for the restriction enzymes *Xba*I and *Xho*I in a nonessential region of the genome. Hence, any vector derived from (but smaller than) wild-type λ may be potentially used as an insertion vector at these sites. In addition, both *Sal*I sites and both *Sst*I sites are also located in a nonessential region of the phage. By cutting at these sites, one can use λ-derived phages as substitution vectors. In fact, sites *Xba*I, *Xho*I, *Sal*I, and *Sst*I can be used in conjunction with the more conventional sites (*Eco*RI, *Hin*dIII, and *Bam*HI) in the different vectors.

Figure 6.1 describes the different steps used for the construction of the λgt cloning vector and its derivatives (Thomas et al., 1974). The linear map at the top of the figure shows the position of the *cos* ends and the location of the six *Eco*RI sites. These sites delineate six restriction fragments labeled A to F. In λgt, the two *Eco*RI sites located at 39.7 and 45.6 kbp on the wild-type genome

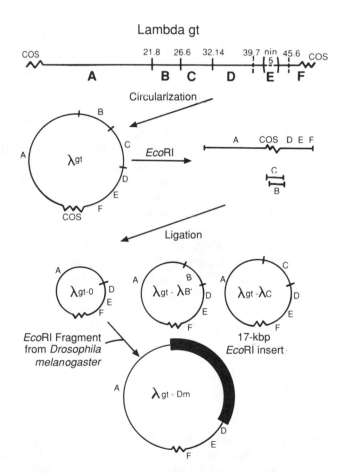

Figure 6.1. *Bacteriophage λgt and its derivatives missing the B and C domains* (from Thomas et al., 1974). On the upper map, the position of the *Eco*RI sites are indicated by vertical bars and kbp; the name of the corresponding fragments is shown in bold letters. The dashed lines correspond to the *nin*5 deletion and the *Eco*RI sites removed. The position of the *cos* sites is also shown. The passenger DNA is shown as a thick line.

have been eliminated. In addition, a deletion (*nin*5) corresponding to 6.1% of the genome was introduced. This deletion removes a strong terminator controlling the expression of the genes located on the right arm of the genomic DNA. It consequently increases the expression of genes *S* and *R* whose products are involved in the lytic pathway. *Eco*RI digestions followed by religation and recircularization of the phage DNA resulted in the construction of vectors λgt-o, λgt-λB′, and λgt-λC. These λ derivatives lack domains B and C, C, or B, respectively. Since domain B consists of 5.5 kbp (i.e., 11.2% of the genomic DNA) and domain C encodes 5.9 kbp (or 12% of the genome), phages carrying both deletions (23.2% of the genome) can still enter the lytic pathway when foreign DNA fragments 1–14 kbp long (i.e., up to 28% of the genome) are cloned into the phage. The *nin*5 deletion in λgt-o removes an additional 6.1% of the phage genome. Hence λgt-o lacks a total of 6.1 + 23.2 = 29.3% of the wild-type genomic DNA. Because this value falls under the lower DNA length limit necessary for phage encapsidation, a progeny will only be obtained if at least 2.2 kbp (and at most 16.9 kbp) of extraneous DNA have been inserted in the phage. This particular feature of λgt-o makes possible the *positive selection* of recombinant phages following cloning and infection of the host cells. Indeed, if both religated phage and recombinant phage could be encapsidated one would obtain plaques from both sets and would have to do further screening to identify the recombinants. It is also interesting to note that fragment C carries the *att, int,* and *cis* genes. Therefore, its deletion prevents the integration of the phage within the host chromosome and abolishes the lysogenic pathway.

At present, over 100 λ derivatives have been constructed. One of these, λWesλB′ (Leder et al., 1977), carries *amber* mutations in genes *W*, *E*, and *S* and is only able to grow on *amber* suppressor recipient strains. Such strains rarely occur in nature, and only a small set of laboratory *E. coli K-12* contain the mutation. Bacteriophage λWesλB′ carries several additional mutations. Fragment C was deleted, while fragment B′ corresponds to the B fragment inadvertently inverted in the construction process (Fig. 6.2). For cloning purposes, the B′ fragment can be removed by *Eco*RI or *Sst*I digestion followed by preparative electrophoresis or sedimentation on sucrose gradients. A foreign DNA fragment flanked by two *Eco*RI sites can then be ligated to the two isolated phage arms and the ligation mixture can be used to transfect an appropriate *amber* suppressor *E. coli* strain. Since direct religation of the two phage arms yields a DNA too small for encapsidation (the C and *nin*5 deletions represent 11.3 + 6.1 = 17.4% of the genome, and removal of the B′ fragment following *Eco*RI digestion takes out an additional 9.8%) only recombinant phages will give rise to plaques.

Figure 6.2 also shows a map of one of the Charon vectors (Blattner et al., 1977, see Appendix 7), which has been widely used in the past. Charon 16A carries *amber* mutations in genes *A* and *B* as well as a substituted *lac*5 fragment from the β-galactosidase *lacZ* gene that contains unique *Eco*RI and *Sst*I sites (Fig. 6.2). This feature makes possible the *color selection* of recombinant phages. Protein synthesis from the *lac*5 gene can be directed by addition of the

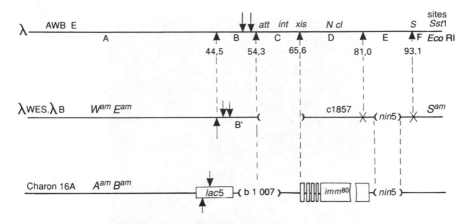

Figure 6.2. *Maps of bacteriophage* λ *and the two derived vectors* λ *WES*λ*B' and Charon 16A.* The positions of the *Sst*I and *Eco*RI sites are shown by arrows on the top and bottom of each map, respectively. Crosses show the *Eco*RI sites that have been eliminated. For the two derivatives, parentheses correspond to deletions and boxes to substitution. The *lac*5 box in Charon 16A is a substitution with the *lac* region from the *E. coli* genome. The *imm*^{80} box and the four smaller boxes on its left-hand side correspond to immunity regions obtained from bacteriophage φ80.

lac operon inducers IPFG or IPTG to the growth medium. The resulting enzyme is capable of hydrolyzing the chromogenic substrate 5-bromo-4-chloro-3-indoyle-β-D-galactoside (X-gal), an operation that generates a non-diffusible blue pigment. However, when a foreign DNA fragment is inserted at the *Eco*RI or *Sst*I sites of *lac*5, no active enzyme is produced, X-gal is not hydrolyzed, and no color reaction is observed. Thus, Charon 16A cleaved with *Eco*RI or *Sst*I, ligated in the presence of extraneous *Eco*RI or *Sst*I-digested DNA, and used to infect host cells plated on a medium supplemented with IPTG and X-gal, will yield blue plaques if the phage has not incorporated the foreign DNA and colorless plaques if the phage is recombinant. It is important to bear in mind that most *E. coli* strains possess β-galactosidase activity. Therefore Charon 16A is usually used to infect recipient strains carrying a *lac* mutation. This selection technique is known as *insertional inactivation;* in the particular case described above, the *lac* gene is inactivated.

The turbid appearance of the plaques obtained following λ infection is the result of the growth of lysogenic cells within the plaque (Fig. 6.3). The protein involved in this phenomenon is the λ phage repressor that negatively regulates the expression of other phage genes and makes lysogenic cells immune to *superinfection* by other phages of the same specificity group. The repressor is encoded by gene *cI*. A large number of insertion vectors where extraneous DNA can be cloned at unique restriction sites within the *cI* gene have been constructed. As a result, recombinant phages yield clear plaques. In many cases, the wild-type λ *cI* gene has been substituted by that of the lambdoid

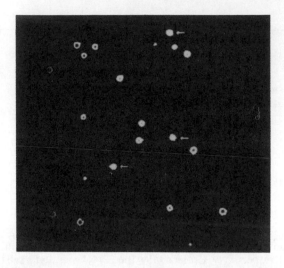

Figure 6.3.   *Turbid vs. clear plaques.* Clear plaques are identified by arrows.

phage 434, which contains unique restriction sites for *Eco*RI and *Hin*dIII. The removal of all other *Eco*RI and *Hin*dIII sites from the genome of the hybrid phages λ*imm*[434] has given rise to a whole series of *insertion cloning vectors* (Murray et al., 1977). A few of these are shown in Fig. 6.4. Such vectors are very useful for the cloning of *Eco*RI or *Hin*dIII DNA fragments of approximately 10 kbp. Recombinant phages grow well and more vigorously than the parental phages due to the clear plaque phenotype and to the fact that their size is optimal for encapsidation.

When such cloning vectors are used to infect *E. coli hflA* strains (for high frequency of lysogeny), no plaques are obtained since large quantities of lysogens resistant to superinfection are produced in these strains. In contrast, recombinant phages have an inactivated λ *cI* repressor and are unable to enter the lysogenic pathway. As a result recombinant phages form plaques on *E. coli hflA,* a property that allows the easy selection of recombinants cloned into vectors such as λgt10 (Nathans and Hogness, 1983).

In some cases, it is also possible to positively select substitution recombinants. A good example is the λWesλB′ derivative λgtWes.T5662 (Davison et al., 1979). In this vector, the substitution fragment of λWesλB′ has been replaced by two identical 1.8-kbp fragments derived from phage T5. These fragments encode gene *A*3, whose product prevents T5 growth on host cells harboring Col1b plasmids. Thus, recombinant phages constructed from λgtWes.T5662 can be positively selected on recipient cells transformed with such plasmids.

Wild-type λ is unable to grow on recipient cells that are lysogenic for phage P2. This property is known as the Spi⁺ phenotype (for sensitivity to P2 interference). In addition, λ bacteriophages carrying deletions in the *red* and *gam*

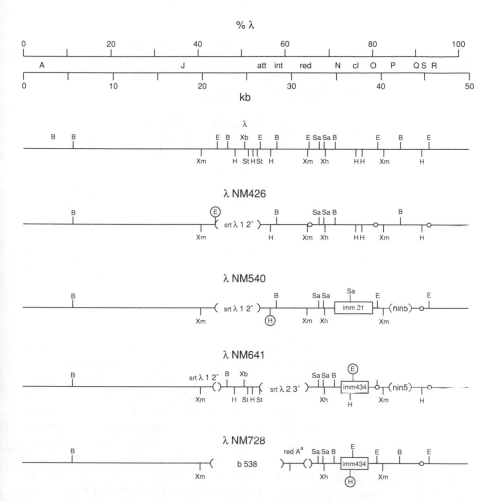

Figure 6.4. *Example of immunity insertion vectors.* The map of wild-type λ is shown on top. Phages NM641 and NM728 are described in the text. Abbreviations for restriction enzyme recognition sites are: B, *Bam*HI; E, *Eco*RI; H, *Hind*III; Sa, *Sal*I; St, *Sst*I; Xh, *Xho*I; Xm, *Xma*I. The sites preferentially used for the cloning of foreign DNA fragments are circled.

genes have a Spi⁻ phenotype. Since both *red* and *gam* are located in nonessential regions of λ, the fragment of phage DNA carrying these genes can be replaced by a foreign DNA and the recombinants can be selected for Spi⁻. Such cloning vectors further carry a *chi* site in the essential part of the genome. *Chi* sites (for crossing-over hotspot instigator) stimulate the formation of multimeric DNA molecules that mature by using the recombination system of the host. (In this case *rec*A⁺ hosts are required.)

Figure 6.5 shows the map of three different substitution vectors useful for the selection of recombinants on the basis of their Spi phenotype. All three have

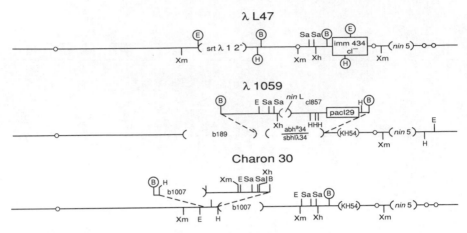

Figure 6.5. *Maps of λL47, λ1059, and Charon 30.*

been constructed for the cloning of large extraneous DNA fragments with *Bam*HI as the main cloning site.

- λ*L47* (Loenen and Brammar, 1980) may be used as an insertion vector at its *Eco*RI and *Hin*dIII sites, It is, however, generally used as a substitution vector through replacement of its central *Bam*HI fragment by the desired foreign DNA. Two deletions increase the *cloning capacity* of λL47 by 8.0 kbp. As a result, extraneous DNA fragments flanked by *Bam*HI sites and ranging between 4.7 and 19.6 kbp may be substituted to the central fragment. In addition, the left arm of the phage carries a *chi* site. When the vector is digested with *Bam*HI, the central fragment may be religated either in its original or the opposite orientation if no attempts are made to specifically remove it. In the latter case, the *red* and *gam* genes become separated from their promoter, which results in a Spi⁻ phenotype and complicates the selection of true recombinants. To circumvent this problem, the λL47 cloning vectors have been modified so that another promoter directs the expression of the *red* and *gam* genes (thus yielding a Spi⁺ phenotype) if the original *Bam*HI fragment religates to the phage arms in its opposite orientation.
- λ*1059* (Karn et al., 1980) has been engineered in such a way that its central *Bam*HI fragment includes the native promoter for *red* and *gam*. Hence, the inversion and religation of the *Bam*HI fragment also yield Spi⁺ phages; λ1059 can accommodate foreign DNA fragments in the 6–24-kbp range. One of the main inconveniences of the vector resides in the fact that only a small proportion of the parental phages survive Spi⁺ selection.
- *Charon 30* (Rimm et al., 1980) is one of the most versatile λ derivatives. It carries two tandem copies of the *Bam*HI replacement fragment, which gives the phage a theoretical cloning capacity of 6–19 kbp. Table 6.1 shows the

Table 6.1 Theoretical Cloning Capacity Of The Charon 30 Phage (From Rimm et al., 1980)

Restriction Enzyme(s) Used	Insert Size (kbp)
BamHI	From 6.1 to 19.1
HindIII	From 0.0 to 11.6
EcoRI	From 4.4 to 17.4
SalI	From 0.0 to 12.1
XhoI	From 0.0 to 11.6
EcoRI + BamHI	From 7.2 to 20.2
HindIII + SalI	From 4.0 to 17.0
HindIII + XhoI	From 4.3 to 17.3
EcoRI + SalI	From 5.9 to 18.9
EcoRI + XhoI	From 6.2 to 19.2
SalI + XhoI	From 0.0 to 12.4

theoretical cloning capacities of Charon 30 digested with one or two restriction enzymes.

Among the latest λ vectors constructed, Charon 34 and 35 are some of the most sophisticated (Fig. 6.6.). These vectors can accept foreign DNA fragments ranging between 9 and 20 kbp, and the recombinant phages can grow on rec^- strains (Loenen and Blatdner, 1983). The central replacement fragment is flanked by two 60-bp *polylinker sites,* which allow the cloning of foreign DNA fragments digested with EcoRI, SstI, XbaI, HindIII, and BamHI. (The SalI site in the right arm may also be used.) The inserted DNA fragments can be excised by digesting the recombinants with SmaI and PstI.

EXPERIMENTS

6.1. **Transfection and Bacterial Lysis** (From Blattner et al., 1977)

Grow an adequate host bacterial strain in 100 ml of NZCYM medium (10 g NZ amine, 5 g NaCl, 5 g yeast extract, 1 g casamino acids, 2 g $MgSO_4 \cdot 7H_2O$ per liter; adjust the pH to 7.5 with NaOH) and prepare four aliquots containing 10^{10} cells. Centrifuge at 4,000g for 10 min at room temperature and resuspend the pellets in 3 ml of SM medium. Infect the cells by adding 5×10^7 phage particles (for the λgt Wes-λB series) or 5×10^8 phage particles (for the Charon series) to the bacteria. Mix by rapid vortexing and incubate for 20 min at 37°C with occasional agitation. Transfer each transfected aliquot into 500 ml of NZCYM medium and grow at 37°C with vigorous agitation for 10–12 h. Add

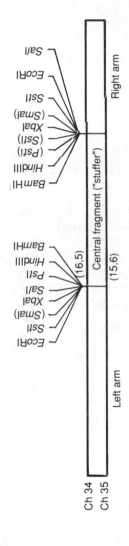

Figure 6.6. *Maps of Charon 34 and Charon 35.* The central fragment (16.5 kbp for Charon 34 and 15.6 kbp for Charon 35) is flanked by a multiple cloning site containing the recognition sequences for several restriction enzymes.

10 ml of chloroform to each culture and further incubate with agitation for 30 min.

6.2. Bacteriophage Purification (From Yamamoto et al., 1970)

To each culture obtained as described above, add 1 μg/ml of DNase and RNase. Incubate the flasks for 30 min at room temperature. (This treatment results in the digestion of the nucleic acids liberated by cell lysis.) Add NaCl and polyethylene glycol (PEG) to a final concentration of 1 M and 10%, respectively. Incubate the flasks on ice for 1 h in order to precipitate the phage particles. Centrifuge the samples at 11,000g for 10 min at 4°C and resuspend the pellet in 8 ml of SM buffer. Add an equal volume of chloroform, mix, and separate the phases by centrifugation at 1,600g for 15 min at 4°C. Recover the aqueous phase containing the phage particles and dissolve 0.5 mg/ml of analytical-grade cesium chloride in it (CsCl). Prepare three CsCl solutions with densities of 1.7, 1.5, and 1.3 in SM medium. (To determine the amount of CsCl to be added, use density tables; for instance, the $d = 1.5$ solution is made up by dissolving 67 g/100 ml in 82 ml of SM buffer.) Carefully layer the solutions by order of decreasing densities in an ultracentrifuge tube in order to generate a CsCl gradient (Fig. 6.7). Layer the cell lysates on top of the gradients and fill the tubes to the rim with mineral oil. Spin the tubes at 30,000 rpm in an SW41 rotor for 3 h at 20°C.

Under these conditions, the phage band sediments at the middle of the $d = 1.5$ zone. Remove the bacteriophage DNA using a syringe and the setup described in Fig. 6.8. Layer the DNA on top of a 5-ml solution of CsCl of $d = 1.5$ poured into a fresh ultracentrifuge tube and spin at 30,000 rpm for 22 h at 20°C. Recover the phage band and dialyze three times against 10 mM Tris-HCl, pH 7.5, 10 mM MgCl$_2$ in order to remove contaminating CsCl.

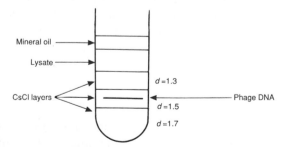

Figure 6.7. *Cesium chloride gradient used in the purification of bacteriophage* λ *DNA.* d = density.

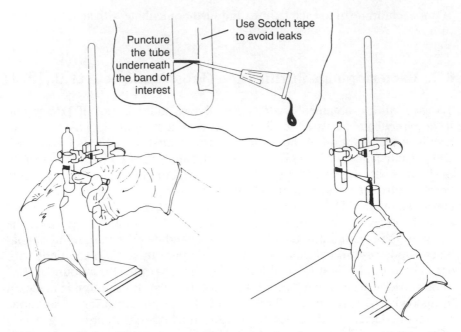

Figure 6.8.   *Recovery of the bacteriophage λ DNA band using a syringe and transfer to a nitrocellulose tube* (from Maniatis et al., 1982).

6.3. Prophage Induction

To verify the lysogeny of the culture, streak the selected lysogenic strain on two NZCYM petri dishes and incubate one at 30°C and the other at 42°C. Colonies should only be obtained on the 30°C dish. Pick a single colony from the latter and use it to inoculate 100 ml of NZCYM medium. Incubate overnight at 30°C with vigorous agitation. At OD_{600} = 0.05 inoculate 500 ml of NZCYM pre-warmed to 30°C. Grow under vigorous agitation at this temperature to an OD_{600} = 0.50 and induce the culture by 15-min incubation at 45°C. Transfer the induced culture to a 38°C water bath and incubate for another 5 h. If the phage carries gene *S*, bacterial lysis will occur spontaneously. Add 10 ml of chloroform to the culture and continue the incubation for 30 min.

6.4. Isolation of Phage λ DNA

To the phage culture obtained as described in the previous section, add EDTA (from a 500 mM, pH 8.0, stock solution) to a final concentration of 20 mM, SDS (from a 20% stock solution) to a final concentration of 0.5%, and protein-ase K to a final concentration of 50 µg/ml. Incubate the lysate for 1 h at 65°C. Precipitate the proteins by addition of one volume of TE-equilibrated phenol.

Separate the phase by 5-min centrifugation at 1,600*g*. Recover the aqueous phase and reextract with one volume of 50% TE-equilibrated phenol, 50% chloroform. Recover the aqueous phase and dialyze to eliminate residual phenol.

6.5. Preparation of Phage Arms

In order to obtain an equimolar preparation of the phage arms, ligate the phage DNA to itself for 2 h at 15°C. Deactivate the DNA ligase by 10-min incubation at 70°C. Next, remove the replacement fragment by digestion with the appropriate restriction endonucleases. For instance, in the case of vector λL47.1, add 1–5 units of *Bam*HI and *Xho*I per μg of phage DNA dissolved in the appropriate digestion buffer. Perform the digestion in a final volume of 500 μl (in order to dilute the ligation buffer) for 12 h at 37°C. The efficiency of the cuts can be verified by running a small aliquot on a 0.8% agarose minigel. Separate the replacement fragment from the heavier phage arms by 16-h centrifugation at 30,000 rpm and 15°C on a 5–40% sucrose gradient prepared in 10 mM Tris-

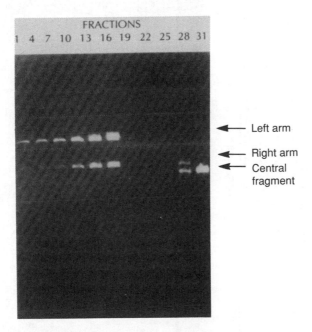

Figure 6.9.   *Isolation of the phage arms by centrifugation on sucrose gradient.* Charon 28 DNA was digested with *Bam*HI and fractionated on a 10–40% sucrose gradient. Aliquots from different fractions in the gradient were resolved by agarose gel electrophoresis and visualized by ethidium bromide staining and UV illumination. Fragments corresponding to the left-hand side (23.5 kbp), the right-hand side (9 kbp), or the central fragment of the phage (6.5 kbp) are labeled with arrows.

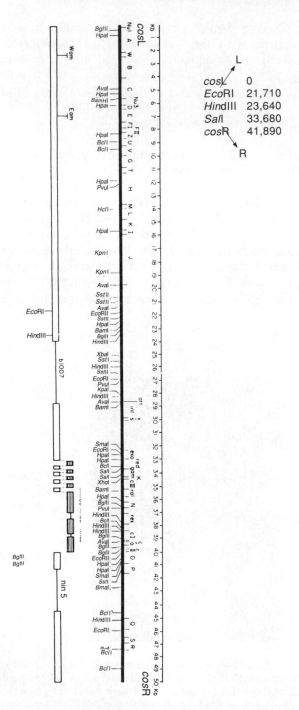

Figure 6.10. *Map of Charon 21A* (from Maniatis et al., 1982) and table giving the length of the restriction fragments from the left-hand side (*cos*L) and right-hand side (*cos*R) *cos* sites.

HCl, pH 8.0; 1 mM EDTA; 1 M NaCl. Up to 100 μg of digested phage DNA may be layered on top of the gradient. At the end of the centrifugation run, fractionate the gradient in 0.5-ml samples and run a small aliquot of each fraction on an agarose gel (Fig 6.9). Pool the fractions corresponding to the phage arms, ethanol precipitate, and resuspend in TE. It is necessary to repeat the whole process with the pooled fractions (digestion, centrifugation on sucrose gradient, and ethanol precipitation) to guarantee the purity of the phage arms.

PROBLEMS

6.1 Cloning Capacity at Different Sites

Using the Charon 21A map and the table giving the length of the fragments cleaved by different restriction enzymes as measured from the left cohesive end—both of which are shown in Fig. 6.10—deduce the cloning capacity when the phage is used:

1. As an insertion vector at site *Eco*RI, *Hin*dIII, or *Sal*I
2. As a substitution vectort at site *Eco*RI-*Hin*dIII, *Hin*dIII-*Sal*I, or *Eco*RI-*Sal*I

6.2 Blue/Clear Plaque Selection

What will be the color of the plaques obtained for wild-type and recombinant bacteriophages in the following cases (justify your answers):

1. The vector encodes the *lac*5 gene and cloning is done by insertion within the gene; the host is *lac*Z$^-$ and petri dishes contain X-gal.
2. The vector encodes the *lac*5 gene and cloning is done by insertion in the operator of this gene; the host is *lac*Z$^+$ and petri dishes contain X-gal.
3. The vector carries the amber suppressors *sup*E and *sup*F and cloning is done by substitution of this region; the host carries an *amber* mutation in the *lac*Z gene; and petri dishes contain X-gal and IPTG.

6.3 Advantages of *Bam*HI Insertion Vectors

In what respect are *Bam*HI insertion vectors advantageous?

Plasmids

Plasmids are small, circular, self-replicating pieces of extrachromosomal DNA that are stably inherited. All the known bacterial plasmids consist of double-stranded DNA. Naturally occurring plasmids encode at least the genes necessary for their replication and segregation in the daughter cells.

Cryptic plasmids are defined as those plasmids that do not induce detectable phenotypic changes. However, most of the known plasmids encode genes that confer specific properties to their host cell (e.g., antibiotic resistance or synthesis, toxin production, resistance to heavy metals, degradation of complex organic ompounds).

As discussed below, naturally occurring plasmids have been improved by genetic engineering to enhance their performance and usefulness as cloning vectors.

7.1. General Properties

- *Size.* Plasmids cover a large span of sizes (from 2 to 200 kbp) depending on their type.
- *Conjugation.* Plasmids larger than 30 kbp frequently encode a set of genes *(tra)* that makes their transfer possible from one cell to the other. A typical example is the F plasmid of *E. coli.*
- *Mobilization.* Plasmids that are too small to encode their own transfer system can be mobilized by a *conjugation plasmid* that coexists in the same cell. For instance, a portion of plasmid ColE1 encodes *mobility proteins* and contains

a site *(nic)* that plays a role in mobilization. However, most plasmids derived from ColE1 (as well as the related pMBL plasmids) have lost these DNA regions. Nevertheless, the gene products necessary for transfer to occur can be provided by other compatible plasmids (such as ColK). In contrast, the only way a plasmid lacking the *nic* site may be mobilized is by cointegration to a conjugation plasmid. This possibility is ruled out in *rec*A mutant hosts.

- *Plasmid copy number.* Every plasmid is maintained at an approximately constant copy number in each cell. Although this number can vary greatly, it is useful to divide gram-negative bacteria plasmids into two groups. The first group consists of those plasmids present at 1–5 copies per chromosome (*low-copy-number* plasmids), while the second includes those plasmids existing at more than 15 copies per chromosome (*high-copy-number* plasmids). There is an inverse relationship between plasmid copy number and plasmid size.

- *Replication.* Only a few copies of plasmids such as pCS101 are present in the cell as a result of a *stringent replication* process. Because the replication of such plasmids is coupled to that of the bacterial genome, they require that the cell actively synthesizes protein and are fully dependent upon DNA polymerase III activity. In contrast, high-copy-number plasmids, such as ColE1, replicate in a *relaxed* fashion that involves DNA polymerase I. Relaxed plasmids are perfectly able to replicate when protein synthesis and chromosomal DNA replication have stopped. This property, which can be used to maximize the recovery of plasmid DNA, is known as *amplification.* If protein synthesis is stopped by addition of the antibiotic chloramphenicol while cell incubation is pursued, the amount of plasmid DNA recovered from the cells can be increased up to 100-fold. (See chapter 8.) Consequently, many genetically engineered cloning vectors contain an origin of replication isolated from relaxed plasmids (e.g., pMB8 carries the pMB1 *ori*).

- *Partitioning.* For each daughter cell to contain plasmid DNA following plasmid replication and cell division, one must postulate a plasmid equivalent of the eukaryotic centromer. Such a segment has been described in plasmid pCS101 and is known as the *par* locus (for partition). However, the majority of cloning vectors are *par⁻*. As a result, most plasmids cannot be maintained in the cell after a certain number of generations unless a continuous *selective pressure* that will kill plasmid-free cells is maintained. This result is generally achieved by supplementing the growth medium with an antibiotic whose action is blocked by a plasmid-encoded gene.

- *Incompatibility.* Two plasmids are *incompatible* when they are unable to stably coexist within a single cell in the absence of selective pressure. In *E. coli,* 25 incompatibility groups have been identified. Since most plasmid cloning vectors have been constructed from a small number of naturally occurring plasmids, incompatibility may be a problem. For instance, ColE1 and pMB1 derivatives, which belong to the same incompatibility group, can coexist in a single cell. (p15A derivatives are also compatible with these plasmids.) Nevertheless, pSC101, F, and RP4 belong to different incompatibility groups, and although any of their derivatives can be stably introduced in a cell transformed with any derivative from the same group, they will not be maintained if the cell harbors, for instance, a ColE1 derivative.

- *Ubiquitous* plasmids are capable of self-transferring in a large number of gram-negative hosts. Such plasmids include the P and W incompatibility groups derivatives.

7.2. Naturally Occurring Plasmids

The term "naturally occurring" is reserved to those plasmids whose construction was not carried out for cloning purposes.

The first cloned DNA fragment was introduced at the *Eco*RI site of the *Salmonella* plasmid pSC101 (Morrow et al., 1974). Plasmid pSC101 (Fig. 7.1) is a derivative of the conjugation plasmid R6-5 and carries the resistance gene for the antibiotic tetracyclin. The *Eco*RI site is located outside of this gene and away from the origin of replication. Plasmid pCS101 is a stringent replication plasmid present at only one or two copies in the cell. Thus, only small amounts of plasmid DNA can be isolated per cell, which is a major impediment in genetic engineering.

This, however, is not the case for ColE1 (Hershfield et al., 1974), a 6.3-kbp

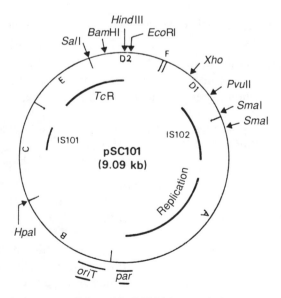

Figure 7.1. *Restriction map of plasmid pSC101* (From Cohen and Chang, 1977). Tc^R, tetracyclin resistance gene; IS101 and IS102, two inserted sequences; *ori*T, sequence involved in mobilization; *par,* partition locus.

medium-copy-number plasmid (20 copies per cell) that has been used exten-sively. ColE1 can be amplified up to 1,000–3,000 copies by addition of chlor-amphenicol to an exponentially growing bacterial culture. Under these con-ditions it may represent up to 50% of the total cellular DNA (Clewell and Helsinki, 1972). ColE1 encodes a gene conferring resistance to colicin E1, an active transport inhibitor. As a result, host cells harboring a ColE1 plasmid are immune to the effect of this substance. The immunity to colicin E1 lysis is the basis for the selection system of plasmid ColE1. The main inconvenient of this technique resides in the fact that colicin resistance arises spontaneously (and at high frequency) in bacteria. Recombinant plasmids obtained by cloning a for-eign DNA fragment at the *Eco*RI or *Sma*I sites located within the colicin E1 gene are unable to protect the cells from lysis by colicin E1 as a result of inser-tional inactivation. In contrast, cells harboring wild-type ColE1 plasmids grow normally on plates supplemented with colicin E1. As a result, recombinant plasmids can be selected on the basis of their inability to grow on colicin plate. This process is known as *negative selection.*

A number of ColE1 derivatives have been developed to facilitate the cloning of extraneous DNA. For instance, pRSF2124 is a ColE1 derivative containing a transposon carrying the β-lactamase gene (an enzyme conferring resistance to ampicillin) and provides a better selection system.

7.3. Characteristics of the Ideal Plasmid

The ideal plasmid for genetic engineering applications should possess the fol-lowing characteristics:

- An easily selectable phenotype
- A large number of unique restriction sites, preferentially in the genes encod-ing selectable phenotypes, to facilitate the cloning and selection of recombinants
- A small size to improve stability, facilitate isolation and transformation, and lower the probability of multiple restriction sites
- A relaxed replication control to allow its purification in large quantities

If a plasmid provides several selectable phenotypic characteristics (e.g., resist-ance to two different antibiotics), the selection process is further simplified. When a DNA fragment of interest is cloned into a unique restriction site located in the gene encoding antibiotic resistance "A", the host cell becomes sensitive to this antibiotic. Therefore, transformed cells can be plated in the presence of antibiotic "B" in order to kill plasmid-free cells and recombinant plasmids can be easily identified through their resistance to drug B and sensi-tivity to drug A.

Plasmid pBR322 (Bolivar et al., 1977a,b) displays all these characteristics and has been widely used as a cloning vector.

7.4. Plasmid pBR322

In an effort to improve the usefulness of naturally occurring plasmids as cloning vectors, genes coding for antibiotic resistance were introduced into ColE1-derived plasmids. This result was achieved both by in vivo transpositions followed by selection of transposed plasmids, and by in vitro insertion of DNA fragments from other plasmids. Figure 7.2 shows the main (known) steps used in the construction of plasmid pBR322.

Plasmid pMB1, which encodes the gene for ampicillin resistance and the *Eco*RI restriction–modification system, was digested by *Eco*RI* and ligated. Following transformation, a colicin E1–resistance, ampicillin-sensitive plasmid (pMB8, 2.6 kbp) was isolated. An *Eco*RI* fragment from pSC101 encoding the tetracycline resistance gene was introduced into the *Eco*RI-linearized pMB8. A recombinant plasmid, pMB9 (5.3 kbp), which carries the tetracycline resistance gene and the ColE1 origin of replication was selected. This plasmid contains a unique *Eco*RI site, unique *Bam*HI and *Sal*I restriction sites in the tetracycline resistance gene, and a unique *Hin*dIII site in its promoter. It proved useful as a cloning vector but was further modified to simplify recombinant selection. The gene encoding ampicillin resistance was introduced by *Tn*3 transposition of pRSF2124 and a plasmid resistant to both tetracyclin and ampicillin was selected and christened pBR312. This plasmid was digested by *Eco*RI* and religated to yield pBR313. The purpose of the latter operation was to eliminate the *Tn*3 sequence in order to (1) generate a unique *Bam*HI site in the tetracycline resistance gene (*Tn*3 also contains a *Bam*HI site) and (2) prevent the *Tn*3-mediated transfer of the ampicillin resistance gene to other epi-

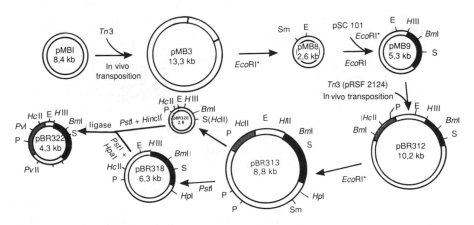

Figure 7.2.   *Main steps involved in the construction of pBR322 from plasmids pMB1, pS101, and pRSF2124* (from Bolivar, 1977). Abbreviations are: E, *Eco*RI; HIII, *Hin*dIII; BmI, *Bam*HI; S, *Sal*I; Sm, *Sma*I; HcII, *Hin*cII; PvI; *Pvu*I; PvII, *Pvu*II; P, *Pst*I; HpI, *Hpa*I.

somes. Finally, two *Pst*I sites were eliminated from pBR313. The resulting plasmid, pBR322, contains a unique *Pst*I site in the ampicillin resistance (β-lactamase) gene. In summary, pBR322 was constructed with:

- The origin of replication of pMB1
- The ampicillin resistance gene from RSF2124
- The tetracyclin resistance gene from pSC101

Figure 7.3 shows a schematic representation of plasmid pBR322 and the relative position of its 20 unique restriction sites. Six of these (*Eco*RV, *Bam*HI, *Sph*I, *Sal*I, *Xma*III, and *Nru*I) are located inside the tetracycline resistance gene; two (*Cla*I and *Hind*III) in the promoter of this gene; and three (*Pst*I, *Pvu*I, and *Sca*I) in the ampicillin resistance gene. Cloning of a DNA fragment in one of these 11 sites will make the recombinant plasmid sensitive to either ampicillin or tetracycline by *insertional inactivation* of the drug resistance. Cloning in any of the other unique restriction sites does not result in a phenotypic change and the presence of a DNA insert can only be ascertained by looking for a size difference between the parental and recombinant plasmids.

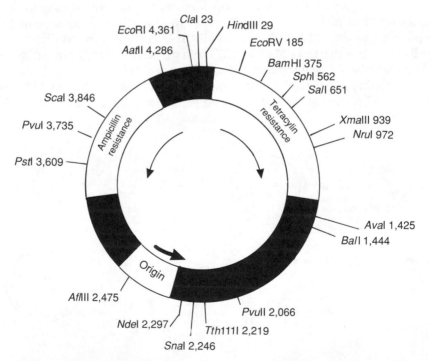

Figure 7.3. *Schematic map of pBR322.* The positions of the unique restriction sites are shown. The *Eco*RI site is used as a reference zero point for the scale. Thin arrows indicate the direction of transcription of the ampicillin and tetracyclin resistance genes. The bold arrow shows the direction of replication of the plasmid DNA from the origin. For a detailed map of pBR322, see Appendix 9.

The sequence of pBR322 was initially published by Sutcliffe (1978) and revised by Peden (1983) and Backman and Boyer (1983) mainly for the addition of the CG doublet at position 526. (See Appendix 8 for the complete sequence.) The DNA sequence consists of 4,362 bp numbered clockwise starting at zero, between the A and T bases of the *Eco*RI site. Every restriction site is numbered at the 5' base located immediately before the cleavage site. (See Appendix 9 for the position of the different restriction sites.)

The origin of replication is located at position 2,534 of the DNA sequence. The tetracycline resistance gene (86 to 1,273) encodes a 396-amino-acid polypeptide. The tetracycline promoter starts at position 44. (Another promoter located in this region initiates transcription in the opposite orientation.) The ampicillin resistance gene (4,146 to 3,297) encodes a 263-amino-acid β-lactamase that is initially synthesized as a precursor protein containing a 23-amino-acid signal sequence at its amino terminal. Transcription initiation of the β-lactamase gene occurs 35 bp upstream of the start codon. (Another β-lactamase promoter initiates transcription 245 bp upstream from the start of the tetracycline gene.)

7.5. Other Useful Plasmids

• *Low-copy-number and secretion plasmids.* In some instances, cloned gene products may be lethal when expressed in large amounts in the host cell (e.g., proteins affecting basic metabolic pathways). It may therefore be useful to clone such genes on plasmids such as pSC101 that are present at only a few copies in the cell. Unfortunately, pSC101 carries only one selectable marker which cannot be used for cloning purposes. Consequently, several derivatives have been constructed. A typical example is pHSG415 (7.1 kbp), which exists at five copies in the cell and encodes three antibiotic resistance genes (streptomycin, kanamycin, and chloramphenicol). Each of these genes contains a unique restriction site suitable for the cloning of extraneous DNA fragments (*Pst*I in *str*, *Hind*III in *kan*, and *Eco*RI in *cat*).

An alternative to the cytoplasmic production of toxic gene products consists in targeting them to the periplasmic space of *E. coli (secretion)* or its growth medium *(excretion)*. Proteins destined for export from the cytoplasm typically contain a short hydrophobic amino terminal extension known as the *signal sequence* or *leader peptide*. After translocation has been initiated, the signal sequence is removed by membrane-bound signal peptidases to yield a *mature* protein. Under certain conditions, it is possible to join the leader peptide of a readily exported protein (e.g., the maltose binding protein, the OmpA protein, or β-lactamase) to the gene of interest. If the resulting *fusion protein* does not fold too rapidly into the cytoplasm and is not inherently incompatible with the export apparatus of the cell, a correctly processed gene product is produced in the periplasmic space or the growth medium.

Some of the most useful secretion vectors, the pINIII-OmpA plasmids, have been developed by Ghrayeb and co-workers (1984) and consist of the *E. coli* OmpA signal peptide followed by a polylinker site where the DNA for the gene of interest can be cloned in one of the three possible reading frames. Transcription is under the control of a modified *lpp* promoter inducible by lactose analogs.

• *pBR322 derivatives.* In 1980, Twigg and Sherrat observed that the deletion of a specific *Hae*II fragment in ColE1 resulted in a five- to sevenfold increase in copy number. When the corresponding *Hae*II fragment was removed from pBR322 (giving rise to plasmid pAT153) a 1.5- to threefold increase in copy number was observed. More importantly, the *Hae*II fragment was found to contain the *nic* site, and its removal prevented plasmid pAT153 from being mobilized. This guaranteed that genes cloned into pAT153 would not be inadvertently transmitted to other bacterial strains.

Early cloning procedures made great use of the restriction enzyme *Eco*RI. However, cloning into the *Eco*RI site of pBR322 did not generate a detectable change in phenotype that would facilitate the selection of recombinant plasmids. As a result, Bolivar and co-workers constructed a series of plasmids suitable for *Eco*RI cloning (Fig. 7.4). Plasmid pBR324, which was derived from pMB9, encodes the genes for the production and immunity to colicin E1. The former gene contains unique sites for the enzymes *Eco*RI and *Sma*I. Thus, insertional inactivation at these sites makes the identification of recombinant plasmids straightforward. Plasmid pBR325 encodes the chloramphenicol resistance gene from phage P1Cm, which also contains an *Eco*RI site. *Eco*RI recombinants can therefore be selected on the basis of their Ap^r Tc^r, and Cm^s phenotype. Plasmid pBR327 was constructed by deletion of a nonessential 1,089-bp sequence in pBR322 and was further modified by exchange of the *Pst*I-*Bam*HI fragment from pBR325 in order to introduce the chloramphenicol resistance gene. The resulting plasmid, pBR328, contains a duplicate of the C-terminal sequence of the tetracycline gene that was found to induce plasmid instability and loss of cloned DNA fragments. Therefore the duplicated region was eliminated with nuclease *Bal*31 to yield the useful cloning vector pBR329.

• *Plasmids allowing for direct positive selection.* pKN80 (Schuman, 1979) contains a DNA fragment from phage Mu that encodes a killing function controlled by the prophage repressor. As a result, this plasmid replicates normally in host strains that are lysogenic for Mu but will kill any other type of host. Insertion of foreign DNA in the *Hpa*I or *Hin*dIII sites of pKN80 inactivates the killing function. Recombinants can therefore be directly selected in strains that are not lysogenic for Mu, on the basis of their resistance to ampicillin (also encoded by pKN80). Plasmid pTR262 (Roberts et al., 1980) encodes a tetracycline resistance gene under the control of a promoter from bacteriophage λ. This plasmid also codes for the λ repressor, which makes it tetracyclin-sensitive under normal circumstances. However, cloning in the *Hin*dIII or *Bcl*I sites, located within the repressor, allows normal transcrip-

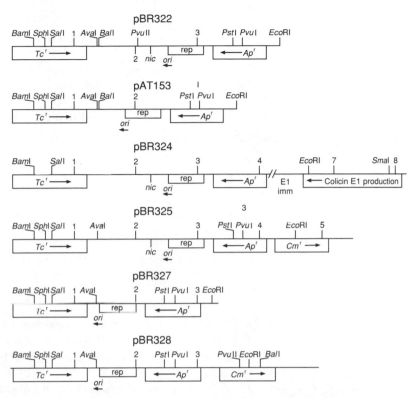

Figure 7.4. *pBR322 and its derivatives* (from different authors). The linear maps start at the *Hin*dIII site and are graduated in 0.2-kbp increments. Arrows indicate the direction of replication (*rep* site) or the direction of transcription of the ampicillin *(Apʳ)*, tetracycline *(Tcʳ)*, and chloramphenicol *(Cmʳ)* resistance genes. The unique restriction sites are shown.

tion of the tetracycline gene. Therefore, recombinants can be selected directly on tetracycline plates.

Visual selection of recombinants is also possible by using plasmids such as pUR2 (Rüther, 1980). On appropriate plates, bacterial colonies containing pUR2 are blue (as is the case with phage M13mp2) while recombinant colonies are not colored.

• *The pUC plasmid series.* pUC plasmids (Viera and Messing, 1982; Norrander et al., 1983; Yanisch-Perron et al., 1985) were constructed using the *Pvu*II-*Eco*RI fragment of pBR322 encoding *Apʳ* and the origin of replication, and an *Hae*II fragment coding for part of the β-galactosidase gene (known as the *α-peptide*). A DNA fragment containing the sequences recognized by several enzymes (*multiple cloning site* or *polylinker*) was isolated from the M13mp phage series and inserted into the *lac* region of the plasmid. In *lac⁻* *E. coli* strains, and on appropriate indicator plates, the pUC plasmids give

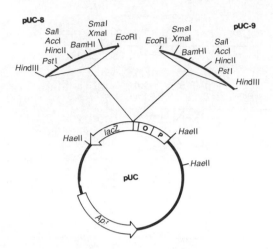

Figure 7.5. *Schematic representation of the pUC cloning vectors.* The multiple cloning site in the *lacZ* gene is shown in the case of pUC8 and pUC9.

blue colonies. However, when a foreign DNA fragment is cloned within the multiple cloning site, the *lac* gene is inactivated and white colonies are obtained. pUC8 (Fig 7.5) is a 2,678-bp plasmid that contains 19 unique restriction sites; six of these are located in the polylinker. Plasmid pUC9 is identical to pUC8 except for the fact that the orientation of the multiple restriction site is inverted. Plasmid pUC18 carries the multiple cloning site from phage M13mp 18, which contains 10 restriction sites. Again pUC19 is identical to pUC18 except for the orientation of the multiple cloning site. (See Appendix 10.)

· *The pEMBL plasmid series.* Plasmids of the pEMBL series (Dente et al., 1983) are relatively small (4 kbp) and very useful for the stable insertion of large DNA fragments. They were constructed by inserting a 1,300-bp *Eco*RI fragment from phage f1 (isolated from pD4; Dotto et al., 1981) in the unique *Nar*I site of every pUC plasmids. The resulting plasmids contain the f1 origin of replication, but, if phage f1 is not present, replication only relies on the ColE1 origin and the average number of copies is about 200 double-stranded plasmids per cell. As is the case for pUC plasmids, the pEMBL series carries the β-lactamase gene and part of the *lacZ* gene in which multiple cloning sites have been introduced (Fig 7.6). By analogy with the pUC plasmids, they were baptized pEMBL8, pEMBL9, pEMBL18, and pEMBL19 depending on the orientation and nature of the multiple cloning site.

Since the pEMBL plasmids contain the f1 origin of replication, infection of host cells harboring pEMBL with phage f1 results in the synthesis of single-stranded plasmids from the double-stranded templates present in the cell. The newly synthesized single-stranded molecules can then be encapsidated

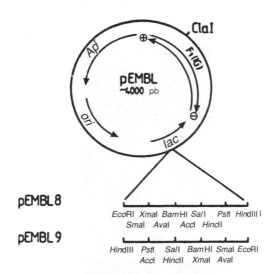

Figure 7.6. *Maps of plasmids pEMBL8 and pEMBL9* (from Dente et al., 1983). The + and − signs at the extremity of the arrows, which show the intergenic (IG) region of f1, correspond to the two possible orientations of the excreted strands on each plasmid. (See text.)

by the proteins produced by f1 and be secreted at the same frequency as f1 DNA if the cell is superinfected. The virions produced in this fashion will contain either f1 DNA or single-stranded pEMBL DNA. Because only one of the plasmid DNA strands is encapsidated and excreted, the pEMBL plasmids contain the f1 intergenic region in one of two possible orientations (labeled + and − in Fig 7.6). For pEMBL$^+$ plasmids, the excreted strand corresponds to the non-coding strand of the α-peptide, while the coding strand is excreted with the pEMBL$^-$ series. Such unusual plasmids that encode both a "normal" (ColE1) and phage-derived origins of replication are known as *phagemids*.

EXPERIMENTS

7.1. Linearization of pBR322 with *Pst*I

One microgram of plasmid pBR322 is digested with 1 μl of *Pst*I, which cuts within the ampicillin resistance gene. The digestion efficiency is verified by DNA electrophoresis. Note that the mobility of the linearized fragment is different from that of the circularized form in Fig. 7.7.

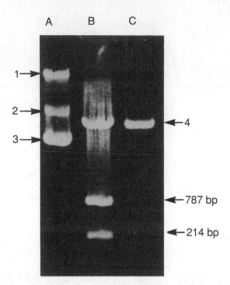

Figure 7.7. *Native and digested pBR322 resolved by agarose minigel electrophoresis.* Lane A: native plasmid. Arrows show the position of the supercoiled monomer (1), the nicked monomer (2), and the supercoiled dimer (3) forms. Lane B: identical sample after *Pst*I digestion. Lane C: identical sample after *Eco*RI digestion. Note that the linearized plasmid DNA migrates faster than its nicked form but slower than its supercoiled conformation.

PROBLEMS

7.1 Characteristics of Plasmids as Cloning Vectors

What are the characteristics of a plasmid that is useful as a cloning vector?

7.2 Advantages of Small Plasmids

What are the advantages of small-size plasmids?

7.3 Directed Cloning

Devise an experimental protocol to insert a *Hin*dIII-*Bam*HI foreign DNA fragment in pBR322. How would you select recombinant plasmids? What is the advantage of this approach?

7.4 The Multiple Cloning Site of pUC18

Using Appendices 2 and 3 identify the restriction enzyme sites (and that of their isoschizomers) present in the pUC18 multiple cloning site:

5′-GAATTCGAGCTCGGTACCCGGGATCCTC-
                               TAGAGTCGACCTGCAGGCATGCAAGCTT-3′

CHAPTER

8

Purification of Plasmid DNA

Hybridization and sequencing experiments require relatively large quantities of purified plasmid DNA encoding the insert of interest. The extraction of plasmid DNA involves three major steps:

- Growth of the bacterium harboring the plasmid of interest and amplification of plasmid DNA
- Cell lysis
- Purification of the plasmid DNA

Several techniques can be used to isolate plasmid DNA. They differ mainly in the way cell lysis is achieved and in the specific steps used to separate plasmid from genomic DNA. The main criteria in selecting a technique are the size of the plasmid to be extracted and the degree of purity desired.

8.1. Bacterial Growth and Plasmid Amplification

An inoculum is prepared with a single colony of bacterial cells proven to harbor the plasmid of interest. The inoculum is used to start medium- to large-scale cultures in the presence of the desired selective pressure (e.g., appropriate antibiotics). These cultures are generally grown with vigorous agitation in rich medium at 37°C.

Relaxed plasmids, such as ColE1 derivatives (e.g., pBR322), can be *ampli-*

fied by addition of chloramphenicol to the growth medium (Clewell, 1972). This antibiotic stops chromosomal DNA replication but does not affect the independent plasmid DNA replication. If the plasmid of interest encodes the gene conferring resistance to chloramphenicol, streptomycin can be used to achieve a similar result.

8.2. Cell Lysis

Bacterial lysis is carried out in a two-step process. First, the cells are incubated with *lysozyme,* an enzyme that affects the structural integrity of the cell wall by attacking the peptidoglycan layer. This treatment induces the formation of spherical cells stripped of most of their outer membrane, the *spheroplasts.* The operation is performed in the presence of sucrose to maintain a constant osmotic pressure and preserve the integrity of the spheroplasts. The solution is buffered in Tris and EDTA since the latter chemical can effectively remove cell-wall components and expose the peptidoglycan layer in gram-negative bacteria such as *E. coli.*

Bacterial lysis is completed by treatment with a detergent. Ionic detergents (e.g., SDS, sarcosyl) liberate the entire cellular DNA. Genomic DNA can then be fragmented while the smaller plasmid DNA is recovered intact due to its specific conformation. This technique is very useful for the extraction of large plasmids. If a non-ionic detergent is used (e.g., Triton X-100, Brij 58), most of the chromosomal DNA remains associated with cellular material. Low-speed centrifugation may then be used to pellet out genomic DNA. The supernatant, or *clarified lysate,* contains prepurified plasmid DNA as well as proteins and other contaminants. This approach is usually used for the purification of small, high-copy-number plasmids.

8.3. Plasmid DNA Purification

When a concentrated solution of *cesium chloride* (CsCl) is centrifuged at high speed ($100,000g$ for 4 h at 15°C), a stable density gradient is formed. If DNA molecules are layered on top of the CsCl solution, they will sediment at a given position in the gradient, depending mainly on their base composition. Ethidium bromide intercalates between the DNA base pairs of molecules with different conformations and, in doing so, decreases their density in the gradient. Since more ethidium bromide can bind to linear and nicked DNA molecules relative to those that are covalently closed or supercoiled, the former conformations have an apparent lower density. Therefore, the supercoiled form of plasmid DNA can be easily separated by ultracentrifugation and DNA banding on CsCl–ethidium bromide gradients (Fig. 8.1).

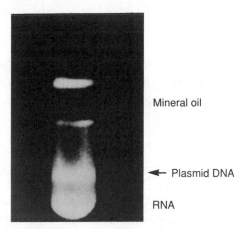

Mineral oil

← Plasmid DNA

RNA

Figure 8.1. *Fractionation of plasmid DNA on a cesium chloride gradient in the presence of ethidium bromide.* The supercoiled plasmid DNA is clearly visible as a large band that sediments on top of the RNA (bottom of the tube). Under the experimental conditions used, chromosomal DNA is virtually absent. A bacterial protein aggregate can be seen under the mineral-oil layer.

The above method yields large quantities of very pure plasmid DNA. However, for many experiments (e.g., restriction-enzyme digestion, transformations), such amounts and purity are not required. Hence, rapid isolation methods *(minipreps)* have been designed. Most of these techniques are based on the alkaline lysis of cells. Treatment with NaOH denatures DNA. Nevertheless, in the $12.0 \le pH \le 12.5$ range, the complementary strands of linear and nicked DNA are separated while supercoiled DNA strands remain associated. Following neutralization, the denatured plasmid DNA (as well as the larger chromosomal DNA) precipitates as insoluble aggregates. Supercoiled plasmid DNA can thus be easily purified by centrifugation and used directly. The alkaline lysis method is also used in some instances to obtain large quantities of pre-purified plasmid DNA before CsCl centrifugation.

In the recent years a number of biotechnology companies have developed kits allowing the rapid recovery of moderate quantities (5–100 μg) of pure plasmid DNA from small cultures. Virtually all these methods involve alkaline lysis as a first step in order to remove most of the chromosomal DNA from the cell lysate. The preparation is then loaded onto a silicon-based matrix that binds plasmid DNA reversibly. After a wash step to remove contaminants, the plasmid DNA is specifically eluted from the column using a TE buffer. These methods, which are fast and easy to set up, have proven a useful alternative to CsCl extractions when only limited quantities of a relatively pure plasmid are required.

EXPERIMENTS

8.1. Isolation of Plasmid DNA From Medium-Scale Cultures (>200 ml) by Ultracentrifugation on Cesium Chloride Gradients in the Presence of Ethidium Bromide (Adapted From Clewell and Helinski, 1969).

Prepare an inoculum by transferring a single colony of the cell harboring the plasmid of interest in 20 ml LB medium supplemented with the appropriate drug(s). Incubate overnight at 37°C with vigorous agitation.

The following day, inoculate 200 ml of LB medium supplemented with the desired antibiotic at a 1:20 ratio. Grow the cells to midexponential phase (OD_{600} = 0.6) at 37°C with vigorous agitation. Use 25 ml of the culture to inoculate 500 ml of LB medium as above. Amplify the plasmid DNA by adding 2.4 ml of a 34 mg/ml stock solution of chloramphenicol made in ethanol. (The final chloramphenicol concentration is 170 μg/ml.) Incubate the flask at 37°C with agitation for at least 12 h. Recover the cells by centrifugation (10,000g for 15 min at 4°C). Resuspend the pellet in 1.5 ml of TES buffer (30 mM Tris-HCl, pH 8; 5 mM EDTA; 25% sucrose), and adjust the volume to 2 ml with TES. Add 250 μl of a freshly prepared solution of lysozyme (10 mg/ml in TES buffer). Incubate on ice for 15 min. Add 200 μl of 0.25 M EDTA and 700 μl of 20% Triton X-100. Incubate on ice for 10 min. The solution should become noticeably more viscous. Spin at 37,000g for 20 min at 4°C. Recover the supernatant (clarified lysate) and adjust the volume to 4 ml with TES. Dissolve 3.65 g of CsCl in the clarified lysate. Add 50–100 μl of a 10 mg/ml solution of ethidium bromide. Transfer the solution to an ultracentrifuge tube and top with mineral oil. Spin in an ultracentrifuge at 45,000 rpm for 36 h at 20°C.

Examine the tube under a UV transilluminator and recover the supercoiled plasmid DNA by piercing the side of the tube with a syringe fitted with a needle under the lower DNA band. (See Fig. 6.8.) Transfer to a clean Eppendorf tube. Remove the ethidium bromide by successive isobutanol extractions. When the pink color has disappeared, dialyze the aqueous phase against a large volume of TE (10 mM Tris-HCl, pH 8.0, 1 mM EDTA) in order to remove the CsCl. Recover the plasmid DNA by ethanol precipitation and redissolve in an adequate buffer. Traces of RNA can be removed by incubation with RNase, followed by phenol/chloroform extraction and ethanol precipitation. Because ethidium bromide is a carcinogen, gloves should be worn in all the steps of the manipulation where this chemical is involved.

8.2. Use of A50 Columns

Low-molecular-weight contaminants (e.g., RNA, detergents, salts, peptides) can be removed by chromatography on A50 agarose columns. Resuspended

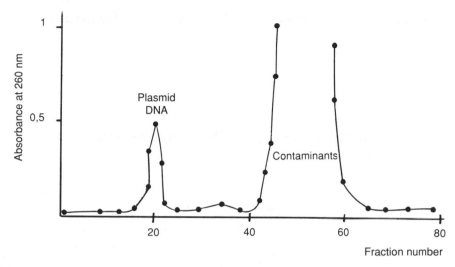

Figure 8.2. *Elution profile of plasmid DNA and low-molecular-weight contaminants following chromatography on a BioRad A50m column.*

the DNA in 500 μl of A50 buffer and 50 μl of 80% glycerol. Layer the solution on top of an A50 column (1 × 20 cm) and elute with 50 mM Tris-HCl, pH 8.0; 0.5 M NaCl; 1 mM EDTA; 1 mM NaN₃ at 1.5 ml/min. Collect fractions at room temperature and measure the absorbance at 260 nm to identify those containing the plasmid DNA (Fig. 8.2).

8.3. Isolation of Plasmid DNA From Small-Volume Cultures by Alkaline Lysis (Adapted From Birnboim and Doly, 1979).

Grow bacteria overnight in 2.5 ml of LB medium supplemented with the appropriate antibiotics. Transfer a 0.5-ml aliquot to an Eppendorf tube and centrifuge at 8,000 rpm for 8 min. Resuspend the pellet in 100 μl of 50 mM glucose; 10 mM EDTA; 25 mM Tris-HCl, pH 8.0; and 2 mg/ml lysozyme. Incubate on ice for 30 min. Add 200 μl of a 0.2% NaOH 1% SDS solution and incubate on ice for 5 min. (The latter solution is prepared fresh by dissolving SDS in TE, adjusting the pH to 12 with 10 N NaOH and to 12.45 with 3 N NaOH; the presence of 50 mM glucose facilitates pH adjustment.) Neutralize with 150 μl of ice-cold 3 M potassium acetate, pH 4.8, and incubate on ice for 20 min. Centrifuge at 10,000 rpm for 5 min; recover the clarified lysate and transfer to a fresh Eppendorf tube. Plasmid DNA can be recovered by ethanol precipitation. Contaminating proteins may be removed by phenol–chloroform extraction. Residual RNA is eliminated by RNase treatment and phenol by ether extraction (Klein et al., 1980). Another rapid method consists of recovering plasmid DNA by isopropanol precipitation following heating at 100°C for 30 sec (Holmes and Quigley, 1981).

PROBLEMS

8.1 Relevance of Plasmid and Chromosomal DNA Separation

In what types of experiments can chromosomal DNA interfere with plasmid DNA? Why is it important to separate the two?

8.2 Plasmid Conformations

What three main conformations can plasmid DNA adopt? (Draw a picture.) What is the impact of these conformations on the physical properties of plasmid DNA?

8.3 Differential Sedimentation of Chromosomal and Plasmid DNA on a CsCl Gradient in the Presence of Ethidium Bromide

Referring to Fig. 8.1, explain why chromosomal DNA is virtually absent in the CsCl gradient, while plasmid DNA is present as a large band. In the presence of ethidium bromide, why does plasmid DNA appear heavier than chromosomal DNA on the CsCl gradient?

8.4 Relative Position of DNA Bands Following Ultracentrifugation on CsCl Gradient in a Fixed-Angle or a Swinging-Bucket Rotor

Using a figure, explain why the resolution of two DNA bands sedimenting close to each other on a CsCl gradient is better on a fixed-angle rotor relative to a swinging-bucket rotor.

8.5 Plasmid DNA Purity

How would you use electrophoresis techniques to verify the purity of a plasmid DNA sample?

CHAPTER

9

Transforming *E. coli* With Plasmid DNA

Bacterial transformation is a natural phenomenon that results in a change in the genetic characteristics of a strain as induced by foreign DNA originating from another strain. Transformation was initially discovered in *Diplococcus pneumoniae* and further demonstrated in *Haemophilus influenzae* and *Bacillus subtilis*. In gram-positive microorganisms, the cell first becomes *competent*—that is, capable of incorporating the extraneous DNA that is initially adsorbed in a double-stranded form. Following completion of this step, an *eclipse phase* occurs. Single-stranded DNA molecules that have penetrated inside the cell and survived degradation by host nucleases insert themselves in the chromosomal DNA by a process known as *homologous recombination*. Foreign DNA stably integrated in the genome replicates, segregates and the proteins it encodes are expressed. The host cell has acquired a new phenotype and is therefore referred to as *transformed*.

Early attempts to transform *E. coli* failed and it was concluded that, in its native state, the cell did not possess the mechanisms required for transformation. In 1970, Mandel and Higa determined that *E. coli* could be transfected by phages P2 and λ following treatment with calcium ions. This method was modified by Cohen et al. (1972), who succeeded in transforming CaCl$_2$-treated *E. coli* cells with plasmid DNA.

The technique developed by Lederberg and Cohen (1974) consists in growing bacteria in a hypotonic solution of calcium chloride at 0 °C. This treatment generates spheroplasts (Henner et al., 1973) and allows extraneous DNA to form hydroxyl-calcium-phosphate bonds. The DNA becomes less susceptible to nuclease degradation and is able to adhere to the cell surface. The complexes can be integrated within the cell by heat shocking the preparation for a small period of time (although heat shock is not necessary for some *E. coli* strains). It is important to bear in mind that the competent state of the cell (which lasts for about 2–3 days) is a direct result of the calcium chloride treatment. If cells are incubated in the presence of calcium chloride for 24 h instead of 30 min, about 20% of the viable cells can be transformed with plasmid DNA (Dagert and Ehrlich, 1979).

The *efficiency of transformation* also depends on the stability of the extraneous DNA following uptake by the cell. Linear DNA molecules are susceptible to degradation by exonucleases and may become unable to circularize and replicate. Hence, it is not surprising that bacterial cells can be transformed 10- to 100-fold more efficiently with supercoiled DNA than with an identical linearized DNA fragment.

A third important parameter in maximizing transformation efficiency is the type of cells used in the experiment. It is important to work with *E. coli* strains carrying mutations in the cell envelope or the restriction–modification system. If plasmid DNA isolated from *E. coli* B is used to transform *E. coli K-12*, the restriction system of the latter will degrade the foreign plasmid DNA since it was not modified by the *K-12* system. The efficiency of transformation will therefore be drastically reduced. In contrast, if plasmid DNA isolated from a given *K-12* strain is used to transform another *K-12* strain, the extraneous DNA will not be degraded and the efficiency of transformation will be much higher. For practical purposes, host cells used to maintain recombinant plasmids contain mutations in their restriction system (*hsd*R⁻ strains).

In addition to the factors discussed above, a large number of parameters also influence the efficiency of transformation. They include the nature and concentration of the divalent cation used to induce a competent state in the cells, the ratio of competent cells to foreign DNA, the pH, the length of the incubation step between DNA and cells, the length of the heat-shock step, etc. For instance, it was recently determined that magnesium ions play an important role in stabilizing foreign DNA during the incorporation process. Therefore, several transformation protocols recommend that a competent state in the cells be induced by using $MgCl_2$ instead of $CaCl_2$.

Using these methods, about 10^5–10^7 transformants can be obtained per μg of pBR322 DNA. This result indicated (1) that only a small portion of the cells can be made competent in such a way that they will stably incorporate foreign DNA and (2) that on the average only one molecule of DNA out of 10,000 will yield a transformant. Specific techniques have been developed to maximize transformation efficiency. The rubidium chloride method developed by Kushner (1978, strain SK 1592) yields 10^7 transformants per μg of recombinant plasmid. In addition, Norgard et al. (1978) have developed a strain (X1776) that gives up 10^8 to transformants per μg of DNA.

In the recent years, a number of alternative techniques have been developed to enhance the efficiency of transformation by plasmid DNA. A typical example is *electroporation* in which *E. coli* cells, thoroughly washed with water to remove conducting salts, are mixed with a small amount of DNA and subjected to a high-voltage electrical pulse. This treatment results (by a-yet-uncharacterized mechanism) in a high uptake of plasmid DNA and, as a consequence, in a large number of transformants.

EXPERIMENTS

9.1. Typical Transformation Protocol

Inoculate 100 ml of LB medium with 1 ml of an overnight culture (e.g., HB101, C600). Grow at 37 °C with vigorous agitation for about 2–3 h to reach a cell concentration of 5×10^7 cells per ml (midexponential-phase cells). Transfer the culture to sterile tubes and spin at 8,000g for 8 min at 4 °C. Discard the supernatant and gently resuspend the pellet in 50 ml of 50 mM $CaCl_2$. Incubate on ice for 20 min and recentrifuge the cells at 8,000g for 8 min at 4 °C. Carefully resuspend the pellet in 2.5 mM of 100 mM $CaCl_2$ and incubate on ice for 12–24 h. The cells will remain competent during this period and the efficiency of transformation will increase with the incubation time. Add the desired quantity of plasmid DNA in TE buffer to a 0.2-ml aliquot of the preparation, mix gently, and incubate on ice for 30 min. Transfer the mixture of 42 °C for 45 sec, place on ice for 2 min, add 0.8 ml of LB medium, and incubate at 37 °C for 30 min (tetracycline) or 1 h (ampicillin) to allow expression of the selectable phenotype. Spin the cells at 8,000g for 30 sec, discard the supernatant, and gently resuspend in 100 μl of fresh LB medium. This mixture can be directly spread on plates containing the required antibiotics.

9.2. Storage of Competent Cells

Competent cells can be stored at -20 °C for up to 3 months or at -80 °C for up to 15 months (Morrison, 1979). For practical purposes, however, competent cells should not be kept for more than 2 months at -80 °C. Once the solution of competent cells has been prepared as described above, add glycerol to 10% final concentration and distribute into 200-μl aliquots in sterile Eppendorf tubes. Freeze the tubes rapidly in a dry ice–ethanol bath and transfer to a cryogenic unit. On the day of use, the competent cells should be thawed gently on ice and transformed as described above.

PROBLEMS

9.1 General Transformation Scheme in Gram-Positive Organisms

Using a figure, explain the mechanism of transformation of gram-positive cells by foreign DNA.

9.2 Frequency of Transformation

How can you experimentally determine the frequency of transformation (give an order of magnitude)? How can the experimenter ensure that the cells have indeed been transformed?

9.3 Calculating the Frequency of Transformation

What is the frequency of transformation if 10 pg of pBR322 yield 100 transformants when 10% of the transformed cells are plated in the presence of the appropriate drug resistance (e.g., ampicillin).

9.4 Calculating the Fraction of Competent Cells

Describe an experiment that could be used to experimentally measure the fraction of competent cells in a preparation.

CHAPTER

10

In Vitro DNA Encapsidation

The theoretical efficiency of bacterial infection by bacteriophage λ is $2 \times 10^{10-}$ *plaque-forming units* (PFU) per μg of DNA. However, when the classic $CaCl_2$ technique is used to introduce naked DNA in the cells, the efficiency is about 10,000 times less. The yield can be improved to 10^8 PFU/μg by using an in vitro encapsidation method that mimics the in vivo process (Sternberg et al., 1977).

Under normal conditions, phage λ *morphogenesis* in *E. coli* proceeds as follows (Fig 10.1): The products of genes *E, B, Nu3*, and *C* assemble in the presence of the host growth protein pE to form a scaffolded prehead (step I). This step involves the *E. coli* molecular chaperones GroEL and GroES, which bind to the scaffolded prehead. Next, the scaffolding is removed through the action of bacterial proteases, and the *A* and *Nu*1 gene products associate with concatemers of λ DNA in the vicinity of the left-hand-side *cos* site. The complex then binds to a specific area of the prehead (step II). In the presence of the FI protein, the phage DNA is encapsidated into the prehead, increasing the size of the latter by about 20% and changing its conformation to an icosahedron (step III). When the head is full, protein D positions itself at the exterior of the capsid, thereby stabilizing the protein structure around the DNA molecules (step IV). During this operation, DNA concatemers are cleaved by the *ter* function of protein A, generating the two free cohesive ends *(cos)* of 12 nucleotides. In this fashion, only one complete λ genomic unit is encapsidated per head. Finally, the tail proteins (I, L, K, G, H, M, V, and U) assemble to form the phage tail in the presence of the *F*II, *W*, and *Z* gene products (step V). A complete phage particle is then obtained.

The in vitro encapsidation process makes use of *cell lysates* isolated from two induced lysogenic strains. The prophages in each strain are identical but carry different mutations that prevent the assembly of mature phage particles (Becker and Gold, 1975; Hohn and Murray, 1977; Hohn, 1979). The first strain (BHB2688) contains an *amber* mutation in gene *E* that maintains all the prohead proteins in a soluble form. The second strain (BHB2690) has an *amber* mutation in gene *D* and accumulates immature proheads that cannot incorporate phage DNA.[1] The lysogens also carry the following mutations: (1) *sam7* to prevent bacterial lysis; (2) *cI*(ts) (within *imm*[434]), which permits growth at 45 °C through inactivation of the repressor; (3) a deletion in region *b* (i.e., *b2* or *b1007*) that affects the attachment site *att* and prevents the excision of prophage DNA following induction, therefore reducing (but not blocking completely) the encapsidation of endogenous phage DNA; and (4) *red3*, a mutation that inactivates the phage recombination system in *rec*A hosts. The latter mutation considerably reduces the amount of recombinants that could arise between the endogenous phage DNA contained in the lysate and the exogenous phage DNA to be encapsidated.

The cell extracts are experimentally prepared by growing the lysogens at 30–32 °C and transferring the cultures to 45 °C in order to induce the lytic cycle. The cells are then incubated at 38–39 °C to accumulate the encapsidation components. When cell extracts isolated from both strains are mixed in the presence of exogenous phage DNA, the mutations are complemented since a complete set of λ proteins is present. In vitro encapsidation of the exogenous phage DNA in the proheads is accomplished by autoassembly of the protein components. In this fashion mature phage particles containing exogenous λ DNA are obtained.

The foreign phage DNA must meet certain requirements to be properly encapsidated. As mentioned previously, it must contain a *cos* site at each of its extremities and its length must range between 78 and 105% of that of wild-type λ. With the in vitro process, DNA concatemers, circular multimers, and linear monomers can all be encapsidated. (Linear monomers cannot be encapsidated in vivo, a property that permits to test of the efficiency of the process.) However, circular monomers cannot be encapsidated in vitro. Therefore, the reaction is carried under conditions that do not favor the formation of such structures. Experimentally, encapsidation is achieved by mixing the exogenous DNA with two buffer solutions supplemented with the polyamines *spermidine* and *putrescine,* the cofactors Mg^{2+} and ATP, and an aliquot of each lysate. A few more complicated experimental protocols have been described and their details can be found in the literature. For instance, the technique developed by Sternberg et al. (1977) involves two types of cellular extracts, while the protocol described by Faber et al. (1978) involves the difficult step of protein A purification (a protein that recognizes a specific site on λ DNA).

---

[1] Other systems make use of non-sense mutations in genes *A* and *E* to achieve similar results (Sternberg et al., 1977).

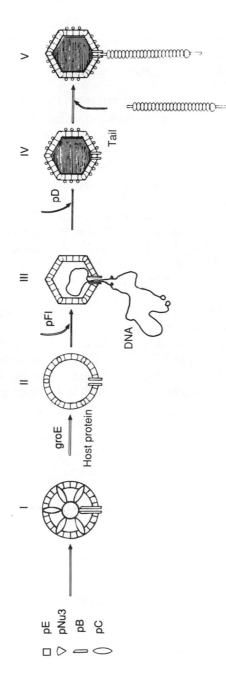

Figure 10.1. *Schematic representation of the different steps involved in bacteriophage λ morphogenesis* (from Hohn, 1979). Gene *E* mutants accumulate all structural proteins with the exception of pE in a soluble form; they are consequently unable to complete the first step of morphogenesis. Gene *D* mutants are blocked at step II. In normal phages, and in the presence of the appropriate DNA molecules, head maturation (steps III and IV) and tail assembly proceed to give a mature phage particle (step V). The prefix pX designates the product of gene X.

EXPERIMENTS

The main advantage of the following experiment lies in the fact that there is no selection on the size of the DNA encapsidated provided that it lies within the limits given above. The main inconvenience is a relatively high background stemming from the encapsidation of endogenous λ.

Prepare lysates from strains BHB2688 and BHB2690 (after testing their genotypes for lysogenicity) as follows. For each strain, inoculate 500 ml of LB prewarmed to 32 °C to OD_{600} = 0.1. Grow the cells at this temperature and under vigorous agitation to OD_{600} = 0.3. Induce the lysogens by shifting the cultures to 45 °C for 15 min. Transfer the flasks to a 38–39 °C water bath and incubate for an additional 2–3 h with vigorous agitation. (Lysogen induction can be verified by adding a few drops of chloroform to a 1-ml aliquot of the culture, which should become clear after 1 min.) Mix the two bacterial cultures, transfer to ice, and pellet the cells by centrifugation at 4,000g for 10 min at 4 °C. Resuspend the pellet in 300 ml of LB and centrifuge as before. Discard the supernatant and resuspend the pellet in 4 ml of complementation buffer (40 mM Tris-HCl, pH 8.0; 1 mM spermidine; 1 mM putrescine; 0.1% β-mercaptoethanol; 7% DMSO) at room temperature. Transfer 50-μl aliquots of the preparation in chilled sterile Eppendorf tubes held on ice and freeze in liquid nitrogen. The lysates may be stored for up to 6 months at −80 °C.

For the encapsidation reaction, dilute 1 μl of DNA (concentration 0.1 to 1 μg/μl) in 5 μl of 66 mM Tris-HCl, pH 7.9, 10 mM $MgCl_2$. Add the appropriate volume of CH buffer and 100 mM ATP (pH 7.5). Mix with a 50-μl aliquot of frozen lysate thawed on ice and incubate at 67 °C for 1 h. At this stage the preparation may be frozen at −80 °C.

PROBLEMS

10.1 Verifying the Genotype of Encapsidation Strains

The genotype of strain BHB2688 is N205 recA λr(λimm434 cIts b2 red3 Eam Sam7). How can you experimentally test the lysogenicity of this strain?

Table 10.1 Encapsidation Efficiency As A Function Of The Amount Of λ DNA To Be Encapsidated

DNA (μg)	PFU	PFU/μg
0.025	2.8×10^5	1.1×10^7
0.100	3.0×10^6	3.0×10^7
0.500	5.5×10^6	1.1×10^7
1.000	3.3×10^7	3.3×10^7

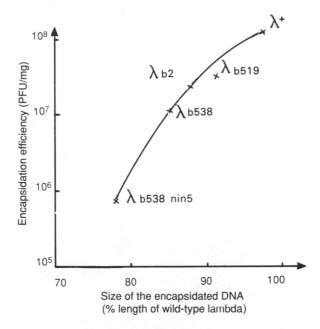

Figure 10.2. In vitro *encapsidation efficiency of λ DNA isolated from different strains* (source Amersham).

10.2 Efficiency of DNA Encapsidation and Amount of Encapsidated DNA

Table 10.1 gives the PFU numbers and the efficiency of encapsidation as a function of the amount of DNA used. What conclusions do you derive from these results?

10.3 Encapsidation Selectivity vs. DNA Size

Figure 10.2 shows the in vitro efficiency of encapsidation for different λ DNA molecules using the Amersham protocol. What are your conclusions?

Part IV
DNA and RNA Preparation

11

Extraction and Characterization of DNA From Tissues

The purpose of the techniques described in this chapter is to obtain long DNA strands (>50 kbp) from tissues. In order to successfully isolate intact DNA, care must be taken to prevent its degradation by endogenous nucleases that become very active upon cell lysis. Hence, the tissues are immediately and rapidly frozen using liquid nitrogen. All the equipment and buffer solutions should be clean and autoclaved.

The general strategy for DNA extraction (Blin and Stafford, 1976) consists of successively eliminating all the cellular components until only DNA remains. The initial step consists of lysing the cells and is followed by protein extraction with phenol–chloroform, dialysis, and RNA digestion (also followed by protein extraction and dialysis).

The tissue containing the DNA of interest is ground into fine powder in the presence of liquid nitrogen by using a chilled mortar and pestle. The ground tissues are resuspended in a lysis buffer containing *EDTA* (which chelates bivalent ions and consequently inhibits DNases, *SDS* (a potent detergent that lyses the cellular membrane and disrupts interactions between nucleic acids and proteins), and *sarkosyl* (which degrades histones and lipids). The combined action of these products is complemented by that of *proteinase K,* a potent protease whose action is favored in the presence of SDS. (See Appendix 11.) Next, proteins are extracted from the solution using a mixture of phenol and metacresol. This step is followed by extractions with mixtures of phenol–chloroform and chloroform–isoamyl alcohol. Chloroform denatures protein surfaces and

Figure 11.1. *Testing the length of extracted DNA by agarose gel electrophoresis.* All three samples are larger than the heaviest fragment of bacteriophage λ DNA digested with *Hin*dIII.

thus facilitates the action of phenol. Isoamyl alcohol acts as an antifoam agent and facilitates the partition of the deproteinated aqueous phase. Organic solvents, sugars, and peptides are eliminated from the aqueous phase by dialysis.

The last step in DNA extraction is the digestion of the endogenous RNA molecules by RNase treatment. The enzyme is removed from the solution by organic solvent extraction and dialysis as described above. As a final step, the size of the extracted DNA is verified by agarose gel electrophoresis (Fig 11.1). A modified protocol for the isolation of DNA from lymphocytes (Gautreau et al., 1983) is described in Appendix 12.

EXPERIMENTS

11.1. Enzyme Preparation

The proteinase K and RNase A (Appendix 13) used in the DNA isolation procedure are first heated to eliminate all traces of DNase activity. Proteinase K (10 mg/ml in ddH₂O) is incubated at 37 °C for 30 min, while RNAse A (10 mg/ ml in 150 mM NaCl) is heated at 80 °C for 15 min.

11.2. DNA Extraction

Grind the desired frozen tissue in a chilled mortar in which a small quantity of liquid nitrogen has been poured. Slowly dissolve the resulting powder in lysis buffer (0.5% SDS, 500 mM sarkosyl-EDTA, pH 8.0) and incubate for 20 h at 50 °C with 100 rpm agitation. Add proteinase K in small aliquots to a final concentration of 100 μg/ml of lysate.

Extract the lysate with one volume of pheno–metacresol–8-hydroxyquinoleine (the organic solution consists of one volume Tris-equilibrated phenol, pH 8.0; 0.14 volume metacresol; 0.12% hydroxyquinoleine). Separate the phases by 10-min centrifugation at 3,000g and discard the top phase. Reextract the bottom phase and the interface with one volume phenol–chloroform (nine parts to one). Spin as above and recover the top aqueous phase. Extract with one volume of chloroform–isoamyl alcohol (24 parts to one), and recover the top aqueous phase after contrifugation. Dialyze against 10 mM Tris-HCl, pH 8.0; 50 mM NaCl; and 10 mM EDTA until the phenol has been totally eliminated ($OD_{270} < 0.05$).

Transfer the dialyzed extract to a fresh flask and treat with RNase A to a final concentration of 100 mg/ml. Incubate for 3 h at 37 °C with gentle agitation. Stop the reaction by addition of SDS (final concentration 0.5%). Add proteinase K (final concentration 100 μg/ml) and incubate for an additional 2 h at the same temperature.

Extract the solution twice with an equal volume of phenol–chloroform and once with an equal volume of chloroform–isoamyl alcohol as described previously. Dialyze the aqueous phase against 10 mM sodium tetraborate, 10 mM EDTA, pH 9.0, until $OD_{270} < 0.05$.

11.3. Spectrophotometric Assay of DNA

The absorbance at 260 nm (in fact, exactly at 254 nm) can be directly correlated to the nucleic acid content of a preparation. A pure, double-stranded DNA solution at a concentration of 1 mg/ml gives an absorbance of 21 at 260 nm. If the ratio between the absorbance at 260 nm to that at 280 nm lies between 1.8 and 2.0, contaminating proteins have been adequately removed from the preparation.

11.4. Chemical Assay of DNA Content

The assay developed by Burton (1956) relies on the dosage of deoxyribose by diphenylamine in the presence of perchloric acid. A set of DNA standard from *E. coli* is used to determine the concentration of the unknown sample. The reaction is followed by spectrophotometry at 600 nm.

11.5. Concentrating DNA Samples

DNA samples may be concentrated by adding one volume is isobutanol and mixing by inversion. The isobutanol, which adsorbs water from the aqueous phase, is removed by using a Pasteur pipette and solubilized in ether that is decanted and evaporated in a nitrogen atmosphere.

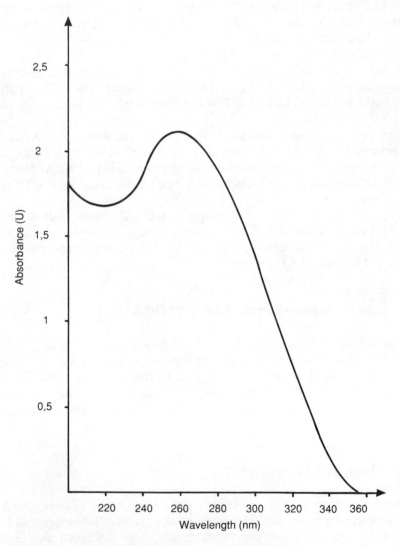

Figure 11.2.   *Absorption spectrum of a placenta DNA solution in the last dialysis buffer.*

11.6. Ethanol Precipitation of DNA

In the presence of sodium acetate, DNA can be precipitated with ethanol as follows: To one volume of DNA solution, add 0.1 volume of 20% sodium acetate and 2 volumes of 200-proof ethanol prechilled at −20 °C. Mix the tube by inversion and centrifuge the precipitate at 8,000g for 10 min. Remove the ethanol with a Pasteur pipette connected to vacuum and dry the pellet by 10-min lyophilization or speed vac or under nitrogen atmosphere. Resuspend in the desired volume of TE buffer (10 mM Tris-HCl, pH 8.0, 1 mM EDTA).

11.7. Storage of DNA Samples

DNA samples can be stored at 4 °C or be stored frozen at −20 °C in TE buffer or as a precipitate in ethanol.

PROBLEMS

11.1 Absorbance Spectrum of a DNA Sample

Figure 11.2 shows the 200–360-nm absorbance scan of a placental DNA solution in its last dialysis buffer. What useful readings should you make on this spectrum? What can you say from these values and the overall aspect of the spectrum?

11.2 RNA Contamination

The absorbance of a liver DNA sample measured at 260 nm gives a concentration of 1.498 mg/g of tissue. However, when the DNA concentration is assayed by the diphenylamine method, a concentration of 0.365 mg of DNA/g of tissue is obtained. How do you explain this discrepancy? Suggest an experimental method that could be used to verify your explanation.

11.3 Total Yield of DNA Extracted From Blood

In general, about 200 μg of DNA can be extracted from 20 ml of blood. Describe a method that would let you determine the average amount of DNA per cell.

12

Extraction of DNA from Gels

For many genetic engineering manipulations (e.g., restriction mapping, cloning, transformation), it is necessary to extract a specific DNA fragment from a gel, whether horizontal or vertical. The techniques described in this chapter are designed to isolate the maximum amount of uncontaminated DNA from a gel slice. The purity of the sample is very important since contaminants in agarose inhibit a number of important enzymatic reactions, such as restriction-enzyme digestion and ligation, whereas the presence of nonpolymerized acrylamide interferes with others (e.g., reverse transcription). All extraction methods include three distinct steps (Yang et al., 1979; Smith, 1980; Chen and Thomas, 1980):

- The localization of the desired DNA fragment through staining with ethidium bromide or autoradiography
- The sampling of the band of interest, which may be performed by physical excision or other methods, as described below
- The purification of the selected DNA fragment form the gel matrix

12.1. DNA Extraction From Polyacrylamide Gels

This technique may be used when the DNA fragment of interest is smaller than 1 kbp and has been resolved on 3.5–5% acrylamide gels. Ammonium acetate is used as an eluting buffer since ammonium ions dissociate contaminants bound to DNA through ionic interactions, whereas the acetate groups convert them into alcohol-soluble salts.

12.2. DNA Extraction From Agarose Gels

Extraction of DNA fragments from agarose gels is complicated by the fact that most commercially available agaroses are contaminated by polysaccharide sulfates that are extracted from the matrix along with the DNA of interest and may inhibit many enzymatic reactions, as mentioned above. In addition, the larger the DNA to be extracted, the lower the yield of the extraction. For instance, DNA fragments larger than 20 kbp are rarely extracted with an efficiency higher than 20%.

One of the most commonly used extraction technique is *electroelution*. The gel slice containing the DNA fragment of interest is transferred to a dialysis bag that is placed in an electroelution cell. When a voltage is applied, the DNA is forced from the agarose into the elution buffer, where it is retained by the dialysis membrane. The DNA can be recovered from this solution by precipitation.

Another possibility is in situ *electroelution*. The position of the band of interest is determined by staining the gel with ethidium bromide and illuminating it with a hand-held UV lamp. A small incision is made with a razor blade downstream from the band (i.e., toward the anode) and a square of *DEAE-cellulose* or *hydroxylapatite* paper is placed in the cut. When electrophoresis resumes, the migration of DNA continues and the fragment of interest becomes trapped on the membrane. The DNA is eluted from the paper by using a NaCl buffer and precipitated with ethanol. This method is very useful for the purification of large DNA molecules such as the arms of bacteriophage λ.

DNA fragments may also be extracted from agarose by dissolving the gel. *Chaotropic agents* such as Ki or $NaClO_4$ can disrupt the hydrogen bonds in agarose (Blin et al., 1975) and the DNA can be separated from the contaminants by chromatography on a hydroxylapatite column. Alternatively, the agarose may be first eliminated by ultracentrifugation on a Ki density gradient and the contaminating Ki may then be removed by dialysis.

Finally, a different type of agarose matrix known as *low-melting-point agarose* (Weislander, 1979) has proven very useful for the extraction of DNA fragments of up to 25 kbp. Low-melting-point agarose contains hydroxyethylated groups that allow agarose to solidify at 30 °C and to melt at 65 °C. Since the latter temperature is lower than the melting temperature of most DNA molecules, intact fragments can be recovered following melting, complexation with quaternary ammonium ions, extraction with isobutanol, and precipitation with sodium salts.

EXPERIMENTS

12.1. Elution of DNA From Acrylamide Gels (Maxam and Gilbert, 1977)

Determine the position of the DNA fragment of interest by staining the gel with ethidium bromide and illuminating it with UV light. Excise the band with a razor blade, place it on a plastic filter, and crush it using the bottom of a round plastic tube. Transfer the fragments to a plastic tube containing 2 ml of elution buffer (500 mM sodium acetate, 10 mM $MgCl_2$, 100 mM EDTA, and 0.1% SDS) and incubate overnight at 37 °C. On the next day, centrifuge the tube at 10,000g for 10 min at room temperature and transfer the supernatant to a fresh tube, taking care to not carry over any of the acrylamide pellet. Reextract the pellet with 0.5 ml of elution buffer and vortex. Spin the tube and recover the supernatant. Combine the two fractions and place on top of a small, siliconized glass-wool column that will retain any small, contaminating acrylamide fragments. Recover the flow-through, ethanol precipitate the DNA, and redissolve it in TE. Reprecipitate the DNA with ethanol in the presence of 0.3 M sodium acetate; spin and dry the pellet under vacuum. Resuspend in a suitable volume of TE.

12.2. Electroelution From Agarose Gels Using DEAE Paper (Banner, 1982)

Localize the band of interest by ethidium bromide treatment and transillumination. Using a razor blade, make a small incision directly downstream of the band of interest. Cut a small square of DEAE paper (0.5 × 0.5 cm), equilibrate the paper in electrophoresis buffer, and insert it in the incision. Resume electrophoresis for 5–30 min at 100 mA. Use a hand-held UV lamp to verify that all the DNA of interest has been transferred to the DEAE paper. Cut the square in small pieces and incubate overnight at 4 °C in a 500 μl microfuge tube containing 200 μl of elution buffer (100 mM Tris-acetate, pH 8.0, 1 M NaCl). The next day, centrifuge the tube for 1 min to pellet the DEAE paper, recover the supernatant, and ethanol precipitate the DNA in the presence of sodium acetate.

12.3. Electroelution Using Dialysis Tubing (McDonell et al., 1977)

Localize the band of interest as above and excise it with a razor blade. Transfer the gel slice to a dialysis tube filled with 0.5 × TBE and place in an electroelution tank. Electroelute at 100 V for 2–3 h. Inverse the polarity and apply the same voltage for 2–3 min in order to liberate the DNA stuck to the inner side of the dialysis tubing. Recover the elution buffer containing the DNA with a Pasteur pipette, wash the tubing and the agarose with 0.5 × TBE, pool the two fractions, and obtain the DNA by ethanol precipitation.

12.4. Elution From Low-Melting-Point Agarose

Pour a low-melting-point gel at the desired agarose concentration and electrophorese the DNA samples as usual. Low-melting-point gels tend to take more time to solidify and are not as mechanically stable as regular agarose gel. They should therefore be handled with more precautions. Localize the DNA band of interest following ethidium bromide staining and illumination with a UV lamp. Excise the band with a razor blade, transfer to a fresh Eppendorf tube, and incubate at 70 °C for 2–5 min. Add enough TE to decrease the agarose concentration to 0.4% or less and estimate the volume. Add an equal volume of Tris-equilibrated phenol, mix vigorously for 1 min, and spin at 12,000 rpm for 5 min. Recover the upper aqueous phase. A thick, white interface should be visible. If desired reextract the phenol interface and the phenol phase by adding one volume of TE and proceeding as above. Pool the aqueous phases and add an equal volume of chloroform–isoamyl alcohol. Repeat the mixing and centrifugation steps. Recover the DNA of interest by ethanol precipitating the aqueous phase in the presence of 0.1 volume of sodium acetate.

CHAPTER
13

Isolation and Characterization of Messenger RNA

Approximately three-fourths of the total eukaryotic RNA consists of *ribosomal RNA* (rRNA), under the form of 28S, 18S, and 5S rRNA. Two other types of RNA molecules, *transfer RNA* (tRNA, 4S) and *messenger RNA* (mRNA), are also present in the cell. With 1–5% of the total RNA pool, mRNA is the least abundant of all RNA species; it is also the most heterogenous in size.

A major impediment to the extraction of mRNA is its high sensitivity to *ribonucleases* that may be endogenous to the organism or accidentally introduced. Thus, very drastic precautions must be taken in order to isolate intact mRNA molecules. All chemicals should be as pure as possible and all solutions and glassware should be autoclaved. All nonautoclavable equipment and solutions should be decontaminated by treatment with a 0.5% solution of diethyl-pyrocarbonate (DEPC, a potent protein denaturing agent). Furthermore, gloves should be worn at all times to prevent the contamination of the solutions by RNases present on the fingers.

Most protocols used in the extraction of mRNA call for the presence of detergents, such as SDS, in order to denature ribonucleases. It is, however, important to bear in mind that ribonucleases may renature if SDS is omitted in a buffer. This is why all RNA solutions are stored under conditions that minimize the activity of ribonucleases (i.e., in the presence of SDS, or in the form of a precipitate in the presence of ethanol at $-20\,°C$).

The extraction of RNA is performed by isolating all the nucleic acids from the cells and eliminating DNA by treatment with *DNase I*. (See Appendix 14.) The mRNA is separated from the total RNA pool by taking advantage of its specific properties.

In general, isolation of nucleic acids from other cellular compounds is carried out by phenol extraction. The cells are initially homogenized in the cold in the presence of a lysis buffer (Kirby, 1968) consisting of a strong detergent (triisopropylnaphtalene sodium sulfonate), a chelating agent, and sodium chloride dissolved in phenol in order to increase the solubility of the detergent. A lysis-buffer-to-tissue ratio of 10:1 is usually sufficient to inhibit the action of ribonucleases.

Contaminating proteins are removed in the cold by using a solution of saturated phenol (to prevent a volume reduction of the aqueous phase) containing metacresol (an antifreeze agent) and 8-hydroxyquinoleine (an antioxidizing agent). Following centrifugation of the sample, three different phases can be distinguished:

- The top aqueous phase contains the nucleic acids.
- The bottom organic phase segregates hydrophobic cellular compounds.
- The interface is composed mainly of denatured proteins.

The top aqueous phase also contains polysaccharides, lipids, and small-molecular-weight compounds. However, most of these are eliminated by precipitating the nucleic acids with two volumes of 200-proof ethanol at -20 °C. Following centrifugation and drying, the pellet is resuspended and the DNA is specifically digested with DNase I.

Although this method works well for the extraction of total cellular RNA, additional precautions must be taken for the extraction of mRNA. For instance, eukaryotic mRNA contain *polyadenylated* [*poly(A)*] *tails* that increase its affinity for denatured proteins. In order to reduce partitioning, NaCl is not included in the lysis buffer. In addition, the extraction is preformed under alkaline conditions by adding 10 mM Tris-HCl, pH 8.5, to the lysis buffer. Finally, proteins are removed with a 1:1 solution of phenol/chloroform that reduces interactions between RNA and proteins at the interface. Alternatively, the phenol–protein interface may be heat treated.

13.1. Isolation of mRNA by Affinity Chromotography

With the exception of histone mRNA, eukaryotic mRNA molecules contain a 20–250-bp-long polyadenylated (polyA) tail at their 3′ end. Thus, they can be separated from other RNA species by affinity chromatography on an *oligo(dT)-cellulose* column (Aviv and Leder, 1972) or a *poly(U)-Sepharose* column (Lindberg and Persson, 1972). The commercially available cellulose affinity matrix contains 40 mg of 25-nucleotides-long oligo(dT) per g of cellulose. The ligand is active under different salt conditions. At neutral pH and high salt concentrations (0.3–0.5 M NaCl or KCl), the poly(A) tails of mRNA bind to the column while other RNA molecules flow through. Following a wash step with the high salt buffer, the mRNA can be recovered by elution with a neutral buffer that does not contain salts (Fig 13.1).

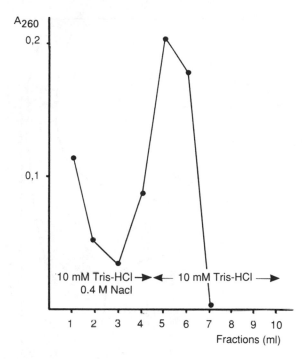

Figure 13.1. *Purification of poly(A)⁺ mRNA on an oligo(dT) column.*

13.2. Analysis of RNA Samples by Sedimentation on Sucrose Gradients

RNA molecules can be separated on sucrose gradients on the basis of their sedimentation coefficient. For example, eukaryotic RNA is readily fractionated on a 5–30% linear sucrose gradient after overnight centrifugation at 80,000*g*. The RNA concentration in the different fractions of the gradient may be directly obtained by measuring the absorbance at 260 nm (Fig. 13.2).

13.3. Assay of mRNA Content by Hybridization to Poly(U)

The poly(A) content of a mRNA preparation can be directly assayed by specific hybridization to tritium-labeled polyuridylate [poly(U)]. Excess poly(U) is eliminated by treatment with ribonuclease A. The isotope content of the hybridized molecules is estimated by precipitation with cold TCA, followed by filtration and scintillation counting. The results are compared to those obtained with standard poly(A). This technique is very sensitive with a detection limit of 1 ng of poly(A).

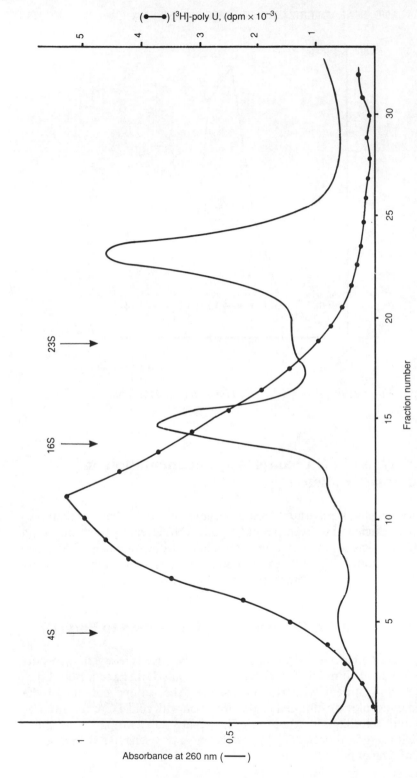

Figure 13.2. *Analysis of whole-liver RNA by sedimentation on a 5–25% sucrose gradient under denaturing conditions.* The direction of the gradient is from left to right. Arrows indicate the positions at which the *E. coli* RNA markers 4S, 16S, and 23S migrate on a similar gradient run in parallel. The RNA sample of interest (45 A_{260} units) was denatured by heating in the presence of 80% DMSO and layered on top of the gradient in 50% DMSO. The absorption at 260 nm was measured for each fraction (solid curve), and the presence of poly(A) in each fraction was determined by hybridization to [³H]poly(U) (dashed curve).

114

EXPERIMENTS

13.1. Isolation of Total Cellular RNA

The technique of Chirgwin et al. (1979) consists of a rapid homogenization of the tissue of interest in a concentrated solution of guanidine cyanate, a strong chaotropic denaturant.

Extract and grind the tissue of interest. Transfer to a homogenizer containing 15 ml of lysis buffer (50 mM sodium citrate, pH 7.0; 4 M guanidine cyanate; 10 mM EDTA; 0.5% sarkosyl, 0.1% β-mercaptoethanol) per g of tissue. Centrifuge the homogenized solution at 5,000 rpm and 10 °C for 10 min in order to pellet cellular debris. Recover the supernatant, add 0.75 volumes of ethanol, and incubate overnight at -20 °C with vigorous agitation.

Centrifuge at 5,000 rpm and 10 °C for 10 min, discard the supernatant, and resuspend the pellet in 0.5 volume of 25 mM sodium citrate, pH 7.0; 7.5 M guanidine hydrochloride; 5 mM DTT. Adjust the pH to 7.0 and heat at 68 °C for 2 min to facilitate the dissolution of the pellet. Adjust the pH to 5.0 by addition of 0.025 volume of 1 M citric acid, and add 0.5 volume of ethanol. Incubate at -20 °C for 4–6 h to precipitate the RNA. Centrifuge at 5,000 rpm at 10 °C for 5 min and discard the supernatant, which contains mainly DNA and low-molecular-weight RNA molecules. Resuspend the pellet as after the first ethanol precipitation, add 0.5 volume of ethanol, and incubate overnight at -20 °C. Recover the RNA pellet by 5-min centrifugation at 5,000 rpm and 10 °C. Wash with 70% ethanol to eliminate salts and respin under the same conditions. Dry the pellet under vacuum and resuspend in 0.25 volume of 10 mM Tris-HCl, pH 7.5, 0.5% SDS. After homogenization, spin at 5,000 rpm and 25 °C for 5 min to pellet insoluble materials. The supernatant contains the RNA.

13.2. Spectrophotometric Assay of RNA

The absorbance of a 1 mg/ml solution of RNA is 25 at 260 nm.

13.3. Affinity Chromotography on Oligo(dT) Columns
(Bantle et al., 1976)

Ethanol precipitate the RNA solution of interest and centrifuge at 7,000 rpm for 30 min. Wash the pellet with 70% ethanol, dry, and resuspend in low-ionic-strength buffer (10 mM Tris-HCl, pH 7.5, 0.5% SDS) to a final RNA concentration of 1 mg/ml. Denature the RNA by heating for 5 min at 60 °C and chill on ice. Add NaCl to a final concentration of 500 mM. Equilibrate an oligo(dT)-cellulose-type T7 column with 10 mM Tris-HCl, pH 7.5; 500 mM NaCl; mM

EDTA; 0.5% SDS and connect the outlet to a detector set at 260 nm and a fraction collector. Load the RNA sample and develop the column at 0.5 ml/min; wash until $A_{260} < 0.05$. The flow-through contains mainly ribosomal RNA. Elute the mRNA with low-ionic-strength buffer and pool the mRNA fractions absorbing at 260 nm.

13.4. Sedimentation of RNA Solutions on Sucrose Gradients

Resuspend the lyophylized RNA in 10 mM Tris-HCl, pH 7.5; 100 mM LiCl; 5 mM EDTA; 0.2% SDS. Denature by heating for 5 min at 37 °C in the presence of 80% dimethylsulfoxide (DMSO). Bring the DMSO concentration to 40% by addition of the previous buffer. Layer the sample on top of a linear 5–20% sucrose gradient and ultracentrifuge for 20 h at 40,000 rpm and 25 °C. On top of a second gradient, layer molecular weight markers (4S tRNA, 16S rRNA, and 23S rRNA, all isolated from *E. coli*). Fractionate the gradients and measure the absorbance of the different fractions at 260 nm.

13.5. Assay of the Poly(A) Content

The assay is performed by specific hybridization to tritium-labeled poly(U) (Bishop et al., 1974). The reaction volume is 1 ml and contains 500 μl of hybridization buffer (20 mM Tris-HCl, pH 7.5, 400 mM NaCl), 0.01–0.02 μCi of [³H]poly(U) at a specific activity of 361 μCi/μmol, and 0.1–1 μg of RNA. Incubate the samples at 30 °C for 30 min and add 25 μl of RNase A (12.5 μg/ml). Incubate for another 30 min at 30 °C in order to degrade the nonhybridized domains of the RNA molecules. Chill the samples to 0 °C; add 3 ml of 8% ice-cold TCA, and 25 μl of yeast tRNA at 2 mg/ml to act as a carrier; and incubate on ice for 15 min. Under these conditions, all RNA molecules smaller than 15–20 bp are precipitated. Filtrate the samples on 0.045-μm Sartorius filters, dry the filters for 1 h at 60 °C, and count the radioactivity of the [³H] poly(U)–Poly(A) hybrids.

PROBLEMS

13.1 RNA Integrity

From Fig. 13.2, what can you say of the integrity of the RNA preparation? Can you make specific comments about the mRNA?

Table 13.1   Fractionation Of A RNA Preparation By Oligo(dT)-Cellulose Chromatography

	RNA concentration (mg)	Poly(A) concentration (μg)
Total RNA	11.69	70.14
Poly(A)$^-$	9.52	4.45
Poly(A)$^+$	0.75	57.97

13.2 Calculating the Concentration of Total Extracted RNA

What is the RNA content of a 20-ml solution of total RNA when 1 ml of a 1:125 dilution absorbs at 0.254 at 260 nm?

13.3 Poly(A) Yield

In an experiment, the RNA and poly(A) contents have been determined by spectrophotometry and titration, respectively. Table 13.1 gives the amounts of RNA and poly(A) in the total extract, the flow-through fraction [poly(A)$^-$], and the bound fraction [poly(A)$^+$] after oligo(dT)-cellulose affinity chromatography. Calculate the percentage of poly(A) present in each fraction. What is the purification factor for mRNA?

13.4 Poly(A) Contamination by Ribosomal RNA

The analysis of fractions enriched in polyadenylated RNA on an agarose gel run under denaturing conditions shows the presence of bands corresponding to 18S and 28S ribosomal RNA. What are your conclusions?

14

RNA Electrophoresis

RNA is a polyanion that migrates toward the anode when placed in an electric field. Under appropriate buffer conditions and agarose (or acrylamide) concentration, the mobility of RNA molecules is roughly proportional to their molecular weight. As discussed in chapter 4 for DNA, the electrophoretic mobility depends upon the charge and the conformation of the molecules resolved. Hence, the accurate determination of the molecular weight of RNA is carried out in denaturing gels. This approach eliminates the migration artifacts induced by the secondary-structure elements in RNA molecules.

The most frequently used denaturing agents for agarose gels are 5 mM methylmercury hydroxide, 2 mM formaldehyde, and 30 mM NaOH (Lehrach et al., 1977). SDS is not particularly efficient at denaturing RNA or dissociating aggregates. The latter problem is usually addressed by heating the samples at 60 °C in the presence of 8 M urea. In the case of acrylamide gels, 98% formamide or 7 M urea is typically used as a denaturing agent.

RNA can be stained with ethidium bromide. Glyoxal denatures RNA under neutral and acidic conditions by reacting with bases (especially with guanosine) and thus prevents its renaturation. Following electrophoresis and staining with acridine orange, RNA molecules can be visualized by UV transillumination at 254 nm.

Specific RNA molecules may be isolated from low-melting-point agarose gels and used in in vitro transcription–translation systems or for the synthesis of complementary DNA.

EXPERIMENTS

14.1. Sample Preparation

Resuspend 10 μg of ethanol-precipitated RNA in 9.3 μl of sterile ddH$_2$O. Add 20 μl of deionized formamide, 6.7 μl of 37% formaldehyde (final concentration 2.2 M), and 4 μl of 12.5 $\times$ sample loading buffer. Heat at 60 °C for 5 min and chill on ice. Add sample buffer, load, and run immediately.

14.2. Methylmercuric Hydroxide Gels (From Bailey and Davidson, 1976)

Prepare a 1% agarose solution in 50 mM sodium borate, 10 mM EDTA, 150 mM boric acid. Heat at 60 °C to melt the agarose. Add 750 μl of a 1 M solution of methylmercuric hydroxide to each 150 ml of agarose solution, and cast the gel. To each RNA sample, add 1 μl of a 0.1 M methylmercuric hydroxide solution, load, and run the gel at 50–100 V. Electrophoresis should be performed in a hood, using a peristaltic pump to recirculate the buffer between the chambers. Stain the gel with 200 ml of ddH$_2$O containing 140 μl of β-mercaptoethanol and 20 μl of a 10 mg/ml ethidium bromide solution.

14.3. Formaldehyde Gels (From Rave et al., 1979)

Dissolve 2 g of agarose in 147 ml of ddH$_2$O and heat to 60 °C. Add 20 ml of a 10 $\times$ solution of 200 mM MOPS, pH 7.0; 50 mM sodium acetate; 10 mM EDTA; and 33 ml of 37% formaldehyde. Cast the gel, load, and run at 50–100 V as above.

14.4. Use of Glyoxal (From McMaster and Carmichael, 1977)

To the RNA samples, add 10 mM of phosphate buffer, pH 7.0. Prepare each sample by mixing 10 μl of the RNA solution, 8 μl of ddH$_2$O and 3 μl of 7 M deionized glyoxal. Heat for 1 h at 50 °C. Add 10 μl of sample buffer (0.05% bromophenol blue, 20% glycerol, 20 mM EDTA). Load and develop the gel as above.

14.5. Preparative Electrophoresis Using Low-Melting-Point Agarose in the Presence of Methylmercuric Hydroxide (From Lemischka et al., 1981)

Run the samples as described in experiment 1, substituting low melting point for regular agarose. After electrophoresis, shake the gel in a 100 mM solution of dithiothreitol for 30 min. Cut 2-mm-wide slices and incubate at 65 °C. Add four volumes of 500 mM ammonium acetate. Phenol extract the RNA from the agarose, centrifuge at 2,000g for 2 min at 4 °C, and precipitate with 70% ethanol in the presence of ammonium acetate.

14.6. Formamide Polyacrylamide Gels

Polyacrylamide gels containing 98% formamide are prepared as follows:

Stock Solution (ml)[a]	Polyacrylamide Concentration (%)			
	5	8	12	20
A	6.7	10.7	16	26.7
B	32.9	28.9	23.6	12.9
C	0.4	0.4	0.4	0.4

[a] Solution A: Dissolve 3.68 g of diethyl barbituric acid in 1 L of deionized formamide. Adjust the pH to 9.0 with HCl. Solution B: acrylamide at 30%. Solution C: ammonium persulfate at 18%.

14.7. Urea Polyacrylamide Gels

Different-percentage polyacrylamide gels containing 7 M urea are prepared as follows:

Stock Solution (ml)[a]	Polyacrylamide Concentration (%)			
	5	8	12	20
Urea (g)	16.8	16.8	16.8	16.8
Solution A	6.7	10.7	16.0	26.7
Solution B	0.8	0.8	0.8	0.8
Solution C	4.0	4.0	4.0	4.0

[a] Solution A: acrylamide at 30%. Solution B: ammonium persulfate at 3%. Solution C: 10 × TBE buffer.

CHAPTER
15

In Vitro Protein Synthesis

The development of in vitro translation systems has opened the way to major advances in the identification and quantification of mRNA molecules by facilitating the study of their translation products. Translation systems are able to synthesize a protein from any purified mRNA molecule. They consist of a mixture of ribosomes, tRNA molecules, tRNA aminoacyl transferases, and cations (K^+ and Mg^{2+}) as well as initiation, elongation, and termination factors. The energy required for the translation of the messenger matrix of interest is provided under the form of ATP and GTP.

15.1. Different Types of In Vitro Translation Systems

Two systems are particularly useful for the in vitro translation of mRNA:

- *Wheat germ extract* (Roberts and Paterson, 1973) is prepared from hard winter wheat following removal of the endosperm. Before use, the system must be supplemented with amino acids, polyamines (spermidine), and energy-generating components (ATP, GTP, phosphocreatine, and phosphocreatine kinase), as well as optimal concenterations of K^+ and Mg^{2+}. Wheat germ extract is an ideal source of ribosomes due to its low level of endogenous mRNA molecules.
- *Rabbit reticulocyte extract* (Pelham and Jackson, 1976) is the most widely used in vitro mRNA translation system. (See Appendix 15 for its preparation.) It is capable of yielding high levels of protein regardless of the molecular

weight of the protein to be synthesized. Purified rabbit reticulocyte lysates contain a significant pool of endogenous mRNA molecules. Thus, exogenous mRNA must be added in sufficient quantity in order to compete with the mRNA already present in the lysate. This problem may be solved by degrading endogenous mRNA molecules with the calcium-dependent enzyme *micrococcus nuclease*. Once hydrolysis is complete, the nuclease is inactivated by chelating all calcium ions with EGTA. At this point exogenous mRNA molecules are added and the translation is allowed to proceed.

Strictly speaking, *Xenopus* oocyte is not a true in vitro translation system. However, it presents the advantage of carrying out the whole protein maturation process and only requires minute amounts of mRNA.

DNA sequences cloned into plasmids under the control of a strong *E. coli* promoter (e.g., *lac, trc, tac,* P_L) may be simultaneously transcribed and translated using an *E. coli S30* extract (Zubay, 1980). The extract is usually prepared from *E. coli* strains containing mutations in proteases that are likely to degrade the newly synthesized proteins (e.g., OmpT and the *lon* gene product). In the classic system, supercoiled plasmid DNA isolated by centrifugation of CsCl gradients is mixed with the extract along with a preparation containing all the components required for the transcription and translation steps (e.g., amino acids, NTPs, tRNAs, appropriate salts, and an energy-regenerating system). The main application of such systems is the synthesis of a small amount of radiolabeled proteins. Consequently, the mix added to the extract usually lacks one amino acid (methionine or leucine), which is added in a radioactive form. Direct transcription and translation of linear DNA fragments is also possible using an extract isolated from *E. coli rec* mutants. Such strains are unable to synthesize the DNA-degrading enzyme exonuclease V, and their extract is thus suitable for the transcription and translation of linear DNA molecules.

15.2. Optimization of the Reaction Conditions

The amount of mRNA used in the translation reaction varies depending on the performance of the extract. However, it generally ranges between 1 and 15 μg for a 20-μl reaction. It is important to make sure that the messenger molecules are free of proteins and polysaccharides that may interfere with the translation process. Furthermore, the mRNA of interest should be dissolved in ddH$_2$O at a concentration of 1–5 μg/μl before addition to the translation system since solvents and chemicals contained in buffer solutions can affect the efficiency of translation. (For instance, alcohols and high-pH buffers are well known to inactivate translation systems.)

The optimum concentrations in magnesium and potassium ions depend upon the mRNA preparation. This is also the case for the reaction temperature (which may be varied from 25 to 37 °C) and the incubation time (from 30 to

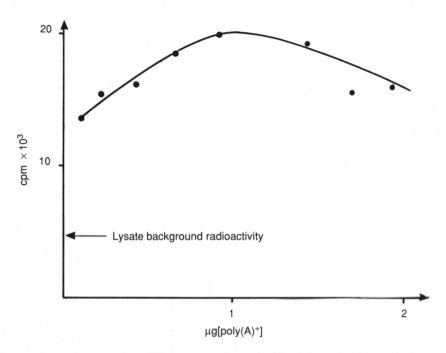

Figure 15.1. *Optimization of the amount of translated poly(A) mRNA in a 20-µl reaction volume for a specific experiment.*

90 min). Hence, optimal conditions must be determined for each experiment by using a test viral mRNA (e.g., that of BMV). Figure 15.1 gives an example of the amount of translated mRNA as a function of the amount of messenger added.

In general, wheat germ extracts must be supplemented with unlabeled amino acids. Rabbit reticulocyte extracts may be purchased in an unfractionated form, which does not require additional amino acids. However, another type of rabbit reticulocyte that has been depleted of its amino acids is commercially available. In this case it is necessary to supplement the reaction with the required amino acids before use.

15.3. Analysis of the Translation Products

In vitro translation systems only synthesize minute amounts of proteins. Thus, the efficiency of the translation process is measured by following the incorporation of radioactive amino acids into growing polypeptide chains. Since ^{35}S is easier to detect than ^{3}H or ^{14}C by autoradiography or fluorography, [^{35}S]-methi-

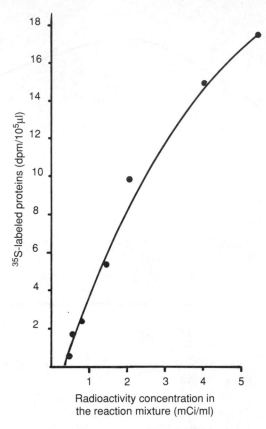

Figure 15.2. *Incorporation of increasing amount of [^{35}S]methionine in the translation products of an mRNA using an unfractionated rabbit reticulocyte system* (source, Amersham).

onine or [^{35}S]-cysteine (which are commercially available at the high specific activity of 1,000 Ci/mmol) is generally used for the labeling of newly synthesized proteins (Fig. 15.2). If the protein studied does not contain any methionine, [^{35}S]-cysteine is the second-best choice.

The amount of radioactivity incorporated in proteins translated from the mRNA of interest can be quantified by precipitating the proteins with trichloracetic acid (TCA, Fig. 15.3). After centrifugation, the supernatant, which contains unincorporated label, is discarded and the protein pellet is counted. To avoid possible artifacts, one must be careful to avoid the formation of nonspecific radioactive precipitates and take into account the labeled tRNA aminoacyls, that are also precipitated by TCA.

In general, translation products are treated with a reducing agent and SDS and resolved by polyacrylamide gel electrophoresis (Fig. 15.4). The identifica-

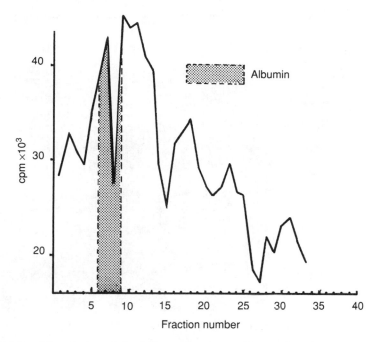

Figure 15.3. *Translation of the different fractions of liver poly(A)⁺ mRNA after sedimentation on a 5–20% sucrose gradient.* Translation of albumin mRNA is shown in darkened area.

tion of particular proteins is then performed by immunoprecipitation and autoradiography (Fig 15.5).

EXPERIMENTS

Denature the mRNA of interest by heating the sample 5 min at 65 °C. The standard reaction medium (25 μl) for the rabbit reticulocyte extract contains 20 mM HEPES, pH 7.4; 1 mM ATP; 0.2 mM GTP; 1 mM DTT; 10 mM creatine phosphatase; 0.3 mM spermidine; 60 μg creatine phosphokinase; 1.2 mM magnesium acetate; 120 mM potassium acetate; 20 μM of all amino acid except methionine; and 2 mCi/ml of [³⁵S]-methionine at a specific activity of 1,240 Ci/mol. The solution is further supplemented with 40 μg/ml of beef liver tRNA, 480 μl/ml of rabbit reticulocyte extract, and 0.5 to 5 μg of the polyadenylated mRNA of interest.

Incubate the mixture at 35 °C for 1 h. As a control of the residual translational activity of the lysate set up the same experiment, omitting the exogenous

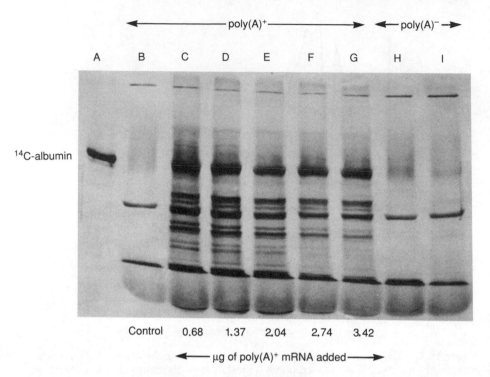

Figure 15.4. *Autoradiogram of a SDS–polyacrylamide gel showing the fractionation of* in vitro *translation products of liver poly(A) mRNA* (rabbit reticulocyte extract, [³⁵S]-methionine 1,000 Ci/mmol). Lane A: [¹⁴C]-albumin marker. Lane B: background translational activity of the reticulocyte extract in the absence of exogenous mRNA. Lanes C to G: translation products using increasing amounts of poly(A)⁺ liver mRNA. Lanes H and I: translation products using poly(A)⁻ liver mRNA. Optimal translation of the poly(A)⁺ liver mRNA is obtained using 1.37 μg.

mRNA. At the end of the incubation period, quantify the amount of newly synthesized proteins as follows. Remove 1 μl from the reaction mixture and transfer onto a 1 cm × 1 cm square of Whatman 3MM paper. Dip into ice-cold TCA at a 10% final concentration. Boil for 10 min. Rinse the filter with water, ethanol, and acetone and allow to air dry. Transfer the filter into 500 μl of Protosol and incubate at 55 °C for 30 min. Count in 10 ml of scintillation fluid in a liquid scintillation counter.

Aliquots of the translation mixture can be analyzed by SDS-PAGE as follows. Mix one volume of sample with an equal volume of loading buffer (125 mM Tris-CHl, pH 6.8, 10% β-mercaptoethanol, 4% SDS, 20% glycerol, 0.006% bromophenol blue, 2 mM L-methionine, 0.04% sodium azide, 5 mM iodoacetate). Denature the proteins by heating at 100 °C for 2–5 min and load on a 8.5–20% gradient polyacrylamide gel.

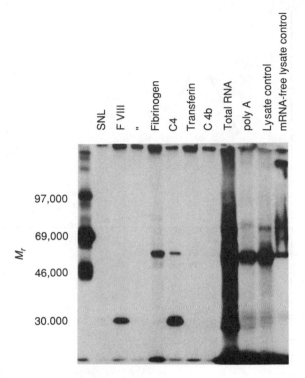

Figure 15.5. *Immunoprecipitation of the C4 fraction of complement and fibrinogen from the total translation products.* Molecular weight markers are shown on the left-hand side.

PROBLEMS

15.1 Effect of the Potassium Concentration on the Efficiency of Translation

The effect of the potassium chloride concentration on the translational activity of a wheat germ lysate and a rabbit reticulocyte lysate is shown in Fig. 15.6. What do you conclude?

15.2 Advantages of [³⁵S]-Methionine

What are the main advantages of [³⁵S]-methionine for measuring the incorporation of radiolabeled amino acids in proteins synthesized in vitro? What additional advantage does Table 15.1 suggest?

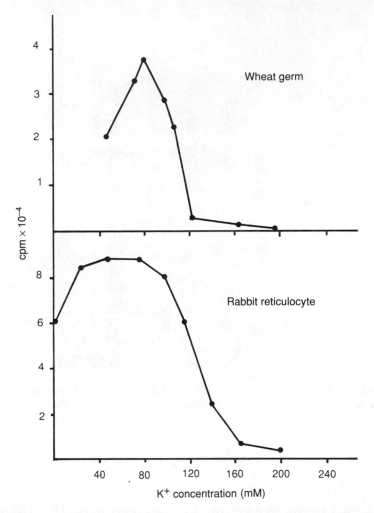

Figure 15.6. *Influence of increasing concentrations of potassium chloride on the incorporation of [³H]-lysine into translated products using a wheat germ extract (top) or rabbit reticulocyte lysate (bottom) system* (from Weber et al., 1977).

15.3 Assaying Different Incorporations

Describe an experiment that will measure the endogenous activity of an in vitro translation system. How can you determine the amount of the protein of interest synthesized by the system? How do you carry out the experiment when the protein of interest is immunoprecipitated?

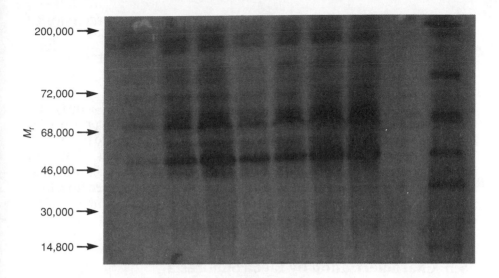

Figure 15.7. *Autoradiogram of the translation products of liver mRNA following sedimentation on a sucrose gradient, immunoprecipitation with antialbumin antibodies, and SDS-polyacrylamide electrophoresis.* Molecular weight markers are shown on the left-hand-side column. The molecular weight of mature albumin is 66,000. The figure shows the SDS-PAGE pattern of immunoprecipitated albumin from the successive fractions of a 5–20% sucrose gradient on the translated poly(A)$^+$ mixture (30μl of antialbumin was used).

Table 15.1 Concentration Of Different Amino Acids In Unfractionated Or Depleted Rabbit Reticulocyte Extracts (From Amersham)

Amino Acid	Concentration (μM)	
	Unfractionated Lysate	Depleted Lysate
L-alanine	264	17
L-arginine	78	4
L-asparagine	52	<1
L-asparatic acid	345	9
L-cysteine	Very low	<1
L-glutamic acid	251	8
L-glutamine	101	<1
Glycine	526	<1
L-histidine	34	5
L-leucine	51	9
L-isoleucine	31	5
L-cystine	115	12
L-methionine	18	6
L-phenylalanine	32	3
L-proline	114	<1
L-serine	120	14
L-threonine	90	6
L-tryptophan	8	<1
L-tyrosine	24	3
L-valine	52	<1

131

15.4 Assaying the Amount of an Immunoprecipitated Protein

In the absence of exogenous RNA, the endogenous translational activity of a rabbit reticulocyte extract corresponds to 4,115 counts per min (cpm) in a 25-μl sample. When exogenous human liver poly(A)$^+$ RNA is added to the extract, 230,210 cpm is measured in the same volume. Among these counts, 26,300 can be immunoprecipitated by using a monoclonal antibody raised against albumin. What percentage of the radioactivity from proteins synthesized from liver mRNA are precipitated by the antialbumin antibodies? Does this amount truly reflect the amount of albumin synthesized by the system?

15.5 Characterization by Electrophoresis

Following in vitro translation of liver poly(A)$^+$ RNA and immunoprecipitation with antialbumin antibodies, radioactive products migrate as shown in Fig. 15.7 on the SDS–polyacrylamide gel. What conclusion can you draw from this gel? Why does the main fraction migrate faster than plasmatic albumin on the gel?

CHAPTER

16

Enriching for Specific mRNA Molecules

Some procedures have been designed to enrich for particular mRNA molecules. Three examples are described below. The first is based on the choice of the tissue from which the RNA will be extracted. The second describes a specific cellular fractionation prior to RNA isolation. The last exemplifies the use of successive oligo(dT) chromatography.

16.1. Tissue Selection

In order to isolate a specific pool of mRNA it is always advantageous to start up with specialized tissues expressing large quantities of the mRNA of interest. A typical example is the extraction of mRNA encoding globins from reticulocytes. In general, the messengers of interest are isolated from tissues that secrete the corresponding proteins. For instance, mRNA coding for steric proteins should be extracted from liver tissues. Furthermore, certain mRNAs are produced at higher levels in a given development stage (e.g., the vitelin sack for α-fetoprotein). Thus, the tissues used in their isolation should be selected accordingly.

16.2. Preparation of Polysomes and Extraction of Polysomal RNA

The mRNA can be defined as the fraction of cellular RNA that sediments with polysomes and can be released by EDTA treatment. In fact, most of the cur-

rently used methods for the extraction of mRNA involve a step in which polysomes are isolated. This operation has the significant advantage of eliminating
contaminating DNA and cellular proteins very early in the purification process. The main inconvenience is that mRNA molecules are also exposed to the
action of ribonucleases in the initial stage of the purification process.

Polysomes are prepared by gently lysing the cells at neutral pH. Typically,
the cells are resuspended in 50 mM Tris-HCl, pH 7.6, containing 150 mM
sucrose, 5–10 mM $MgCl_2$, and 10–50 mM KCl, in order to maintain ribosome
integrity. The cytoplasmic membrane is disrupted by treatment with a detergent (e.g., SDS at a 0.01–0.3% concentration) in the presence of diethylpyro-
carbonate (or heparin) to inhibit ribonucleases.

Following lysis, cell debris is removed by centrifugation at 12,000g for 15
min and the polysomes are purified by sedimentation on sucrose gradients.
Polysomes of different sizes can be separated by using continuous sucrose gra-

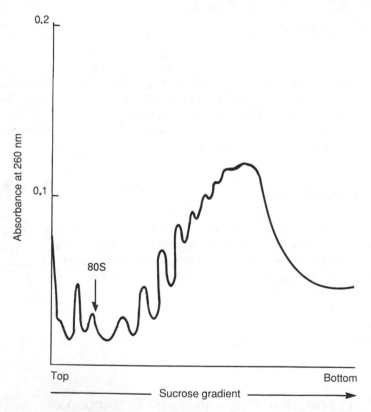

Figure 16.1. *Sedimentation profile of polysomes on a linear sucrose gradient.* Two A_{260}
units of a polysome suspension were layered on top of a 15–50% sucrose gradient in 50
mM Tris-HCl, pH 7.5; 2.5 mM NaCl; 5 mM $MgCl_2$; and 0.1 mg/ml heparin. The absorbance of successive fractions was measured at 260 nm following 90-min centrifugation
at 40,000 rpm. The arrow shows the position of 80S monosomes.

dients (Fig. 16.1). Typically, polysome suspensions are opalescent and yellow and can be assayed spectrophotometrically at 260 nm.

The mRNA of interest is separated from the polysomes by treating the appropriate gradient fractions with 0.5% SDS and 10 mM EDTA, pH 9.0. Contaminating proteins are removed by classic phenol–chloroform extraction.

16.3. Affinity Chromatography on Oligo(dT) Columns

Further purification of the mRNA isolated as described in chapter 13 may be achieved by rechromatographing the poly(A)$^+$ fraction. Figure 16.2 shows the example of a RNA preparation isolated from liver and rechromatographed on an oligo(dT) column. Inspection of the absorption profile shows that this preparation remains contaminated by 5, 18, and 25S RNA. The translation activity of the different gradient fractions (for an identical amount of RNA) is also shown on the same graph. Maximum translational activity arises around 20S; however, due to ribosomal RNA contamination, the peak activity probably occurs in the 18S region of the gradient.

Albumin mRNA sediments at about 18 to 19S in the gradient. Analysis of the translation products by SDS-PAGE indicated that the amount of albumin synthesized progressively increases with increasing gradient concentrations and is maximum between fractions 11 and 12.

EXPERIMENTS

16.1. Polysome Isolation (From Schimke et al., 1974)

Prepare the lysis buffer (20 mM Tris-HCl, pH 7.5; 100 mM KCl; 40 mM NaCl; 7.5 mM MgCL$_2$; 0.25 M sucrose; 1.25 mg/ml heparin; 12 μg/ml cycloheximide; 10 mM β-mercaptoethanol) and chill to 0 °C. (All subsequent operations are performed at 4 °C.) Mix each gram of tissue of interest with 3 ml of lysis buffer supplemented with 1.8 mg/ml of yeast tRNA in order to saturate the RNases. Homogenize and add 150 μl of a detergent solution containing 10% Triton X-100 and 10% sodium deoxycholate. Spin the sample at 11,000 rpm for 10 min, recover the supernatant, and add 2 mg/ml of heparin and 0.055 volume of the detergent solution.

After 10-min agitation at 0 °C, layer 2.5 ml of the supernatant on top of a discontinuous sucrose gradient consisting of a 1-ml layer of 0.5 M sucrose in 25 mM Tris-HCl (pH 7.5), 25 mM NaCl, 5 mM MgCl$_2$, 0.5 mg/ml heparin; a 22-ml layer of 1 M sucrose in the same buffer; and a 6-ml layer of 2.5 M sucrose in the same buffer. Centrifuge at 25,000 rpm at 2 °C for 4 h. Recover the polysome band at the interface between the 1 and 2.5 M sucrose layers. Resuspend

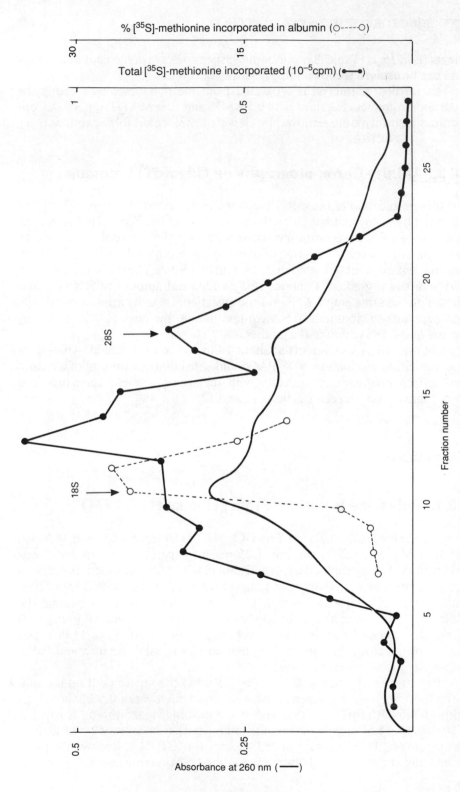

% [^{35}S]-methionine incorporated in albumin (O----O)

Total [^{35}S]-methionine incorporated (10^{-5} cpm) (●—●)

Absorbance at 260 nm (——)

Fraction number

28S

18S

Figure 16.2. *Second fractionation of liver poly(A)$^+$ mRNA on sucrose gradient.* (Compare with Fig. 13.2).

the polysomes in 50 mM Tris-HCl (pH 7.5), 2.5 mM NaCl, 5 mM $MgCl_2$, and 0.1 mg/ml heparin. Analyze the polysome fraction on a 15–50% continuous sucrose gradient as shown in Fig. 16.1.

16.2. Extraction of Polysomal RNA

Rapidly mix the polysome suspension with five to 10 volumes of 20 mM Tris-HCl (pH 7.5), 0.1 M NaCl, 2 mM EDTA, and 1% SDS. Add proteinase K to a 2 mg/ml final concentration and incubate for 1 h at 30–37 °C with continuous agitation. Add 0.5 ml of phenol saturated with the above buffer and agitate vigorously for 10 min. Add 0.5 volume of chloroform and centrifuge at 5,000 rpm for 10 min. Recover the aqueous phase, add SDS to a final concentration of 0.5%, and repeat the phenol–chloroform extraction twice. Finally, add 0.1 volume of 2 M NaCl and two volumes of 100% ethanol to the aqueous phase. Precipitate the RNA at −20 °C and recover the pellet by centrifugation.

PROBLEMS

16.1 Polysome Integrity

What can you say about the integrity of the polysome suspension shown in Fig. 16.1?

16.2 Artifacts in the Sucrose Gradient

Knowing that the RNA layered on top of the gradient shown in Fig. 16.2 contained 50% DMSO, how do you explain the translational activity of the 28S region of the gradient?

16.3 Purification of Albumin mRNA by Second Affinity Chromatography of the Poly(A)$^+$ Fraction

The translation of albumin mRNA represents 5% of the total activity after one affinity chromatography step. When the RNA pool is rechromatographed on the same column, albumin mRNA represents 25% of the total activity. What is the purification factor obtained by subjecting the poly(A)$^+$RNA sample to a second chromatography step?

CHAPTER
17

Complementary DNA Synthesis

Double-stranded *complementary DNA* (cDNA) is obtained from purified mRNA using three successive enzymatic reactions, as illustrated in Fig. 17.1 (Efstratiadis et al., 1976). The enzymes involved in the process are:

* *Reverse transcriptase,* which synthesizes single-stranded cDNA molecules from the mRNA template
* *DNA polymerase,* which generates double-stranded cDNA from the single-stranded cDNA template
* *S1 nuclease* (Appendix 16), which trims off the single-stranded domains of the cDNA molecule unprotected by base paring

17.1. Synthesis of cDNA From Poly(A)$^+$ mRNA

Reverse transcriptase (Appendix 17) does not transcribe all mRNA molecules with the same efficiency. For some messengers, secondary structures from extended-base-paired regions prevent the enzyme from progressing efficiently along the chain. Fortunately, this problem can be (at least partially) avoided by denaturing the mRNA molecule of interest with methylmercury hydroxide.

Reverse transcriptase is a viral enzyme isolated from cells infected with the avian myelobastosis virus (AMV). It is capable of synthesizing DNA (known as complementary DNA or *cDNA*) from an mRNA template. In order to initiate transcription, reverse transcriptase requires the presence of a short oligo-nucleotide *primer* displaying a free 3'-OH group. This is experimentally

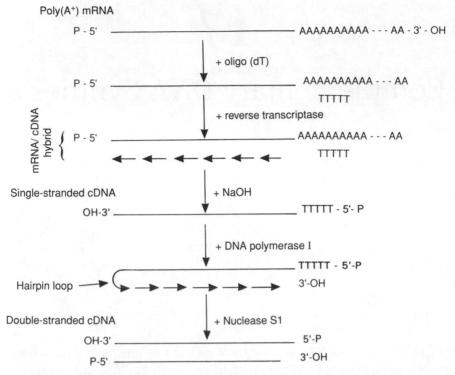

Figure 17.1. *Synthesis of double-stranded cDNA from mRNA.* Single-stranded cDNA is synthesized from the mRNA template using reverse transcriptase; double-stranded cDNA is obtained from single-stranded cDNA using DNA polymerase I and the 5′ hairpin loop is eliminated by treatment with S1 nuclease. (See text for details.)

achieved by mixing the mRNA of interest with a 12–18-bp-long oligo(dT) oligonucleotide. The operation results in the formation of base pairs between the oligonucleotide and the poly(A) tail present at the 3′ end of most mRNA molecules. The oligonucleotide is said to have *annealed* (or *hybridized*) to the mRNA. In doing so, it offers a 3′-OH group suitable for the action of reverse transcriptase. If the mRNA of interest does not contain a poly(A) tail, *random primers* may be used to achieve a similar result.

Optimal conditions for reverse transcriptase activity are very difficult to determine. The activity of the enzyme varies with the degree of purity of the preparation as well as the nature of the mRNA to be transcribed (Retzel et al., 1980). Thus, it is necessary to determine the optimal conditions for each lot of reverse transcriptase and mRNA sample. Some of the important parameters in the reaction are the ratio of enzyme to substrate (the concentration of the latter should be high) and the concentrations of mono- and divalent cations in the buffer solution.

A major impediment to the use of reverse transcriptase is the potential presence of contaminating ribonucleases (particularly ribonuclease H), which can rapidly degrade the mRNA of interest. Thus, all the equipment and solutions used in the transcription reaction must be autoclaved, and the mRNA matrix must be purified by one or two oligo(dT) chromatography steps, as described previously. The amount of contaminants in reverse transcriptase preparations varies from lot to lot although the quality of commercially available preparations has significantly improved in the past few years. Nevertheless, the enyzme may be further purified if necessary. Alternatively, the reaction mix may be treated with RNase inhibitors (e.g., RNasin or vanadyl–ribonucleoside complexes) to improve the yield of the reaction.

The efficiency of transcription can be assessed quantitatively or qualitatively. The quantitative approach consists in measuring the amount of radioactive precursor incorporated into growing complementary cDNA strands. This is experimentally achieved by TCA-precipitating small aliquots of the reaction mixture at different time points and counting. In general, about 20–30% of the initial mRNA mass is transcribed as cDNA at the end of the reaction.

The qualitative approach is performed by resolving radioactively labeled aliquots on alkaline agarose gels (if the cDNA molecules are larger than 1 kbp) or 5% polyacrylamide gels containing 98% formamide (cDNA < 1 kbp) and subjecting them to autoradiography. Figure 17.2 shows that a mixture of cDNA molecules, heterogenous in size, is obtained by transcribing total RNA. This is due to the fact that, in addition to the large cDNA molecules corresponding to the mRNA templates, a large amount of prematurely terminated shorter products are generated by reverse transcriptase.

17.2. Synthesis of Double-Stranded DNA

The molecules obtained in the first step of cDNA synthesis are *hybrids* between mRNA and cDNA. Since the second and third reaction steps require single-stranded cDNA, the annealed mRNA is degraded by a NaOH treatment that does not affect cDNA integrity.

The synthesis of double-stranded cDNA from single-stranded cDNA is the result of an enzymatic reaction that also requires the presence of a primer. Such a sequence is fortunately directly generated by reverse transcriptase in the form of a *hairpin loop* at the 3' end of the cDNA molecule (Fig. 17.1; Rougeon and Mach, 1976). The resulting double-stranded cDNA region is a suitable primer for the synthesis of the complementary strand by a DNA polymerase. In general, the *E. coli* DNA polymerase I is used in the second reaction step (Wickens et al., 1978). The critical parameters for the synthesis of the second cDNA strand are the reaction temperature (which should be maintained at 15 °C), the incubation time (at least 4 h), the nature and concentration of the monovalent ion used in the buffer, and the concentration of DNA polymerase (which must be chosen to suppress the nucleolytic activity associated with the enzyme).

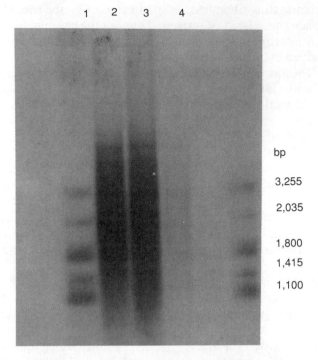

Figure 17.2. *Autoradiogram of a 2% agarose gel showing the heterogenous nature of single-stranded cDNA.* The cDNA was obtained using total poly(A)$^+$ mRNA (lane 1), poly(A)$^+$ mRNA following fractionation on a 5–20% sucrose gradient (lane 2), poly(A)$^-$ mRNA (lane 3), or the wash products of an oligo(dT) cellulose type 2 column (lane 4). Molecular weight markers are shown.

Synthesis of double-stranded cDNA may also be achieved by using the *Klenow fragment* of DNA polymerase I, which is devoid of 5′ exonuclease activity. Alternative enzymes include *T4 DNA polymerase,* and the polymerase activity of reverse transcriptase (Ullrich et al., 1977). The use of the latter is described in the Experiments section of this chapter. It has the main advantage of suppressing an intermediate purification step since both the first and second reactions can be performed in the same tube.

The progress of double-stranded cDNA synthesis can also be assayed by radiolabeling. In the first reaction, the growth of the single-stranded cDNA chain is usually followed by measuring the incorporation of ^{32}P precursors. As a consequence, the evolution of the second reaction is determined by assaying the amount of ^{3}H precursors incorporated in the growing cDNA strand using TCA precipitation and scintillation counting.

Other techniques have been developed to measure the degree of complementary strand synthesis. They include measuring the increased resistance to S1 nuclease digestion (under specific conditions single-stranded DNA is sus-

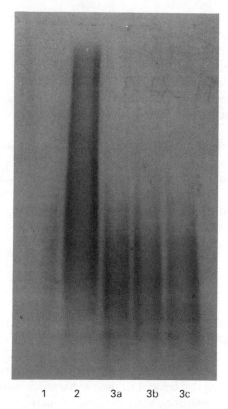

1 2 3a 3b 3c

Figure 17.3. *Autoradiogram of a denaturing 1.4% agarose gel showing the migration pattern of different cDNA molecules.* Lane 1: monocatemeric cDNA synthesized from total mRNA. Lane 2: bicatemeric cDNA. Lane 3: bicatemeric cDNA following S1 nuclease digestion with 1 unit (3a), 4 units (3b), or 6 units of enzyme (3c). Note that monocatemeric cDNA and bicatemeric cDNA treated with S1 nuclease migrate at about the same position while untreated bicatemeric cDNA are twice as heavy.

ceptible to degradation by this enzyme while double-stranded DNA is not) and the differential pattern of electrophoretic mobility on a denaturing gel. (The initial single-stranded cDNA and its newly synthesized complementary strand are covalently linked through the loop and thus migrate more slowly than single-stranded DNA under denaturing conditions (Fig. 17.3).

17.3. S1 Nuclease Digestion

Upon completion of the second reaction, a double-stranded cDNA molecule in which both strands are covalently linked is obtained (Fig. 17.1). Because the hairpin connecting the two strands is composed of single-stranded DNA, it can be digested by *S1 nuclease* under appropriate conditions. (See below.) Since all single-stranded DNA is susceptible to S1 nuclease degradation, the enzyme will also hydrolyze all the cDNA generated in the first reaction for which a protective second cDNA strand has not been synthesized.

S1 nulcease is used under high salt concentrations (300 mM) in order to prevent the hydrolysis of double-stranded DNA. Although an excess enzyme is always used experimentally, it is wise to titrate each preparation of S1 nuclease to guarantee that the double-stranded cDNA is resistant to the concentration of enzyme chosen.

Following S1 nuclease treatment, the cDNA strands will migrate independently on denaturing gels and their electrophoretic pattern will be virtually indistinguishable from that of the single-stranded cDNA obtained in the first reaction (Fig. 17.3). It is important to keep in mind that, regardless of the reaction conditions used, only a small percentage of the original mRNA will be synthesized as double-stranded cDNA.

17.4. Selection of cDNA According to Size

Following digestion by S1 nuclease, a large number of double-stranded cDNA molecules are incomplete. In fact, even the cDNA molecules obtained from a single type of mRNA are heterogeneous with respect to size (i.e., do not correspond to a full-length transcript of the template mRNA). Because small cDNA molecules are easier to clone than large ones, it is important to eliminate the small cDNA molecules present in the preparation before the cloning step. This operation is carried out by fractionating the newly synthesized cDNA on a sucrose gradient in order to eliminate the molecules that sediment slowly (Humphries et al., 1977).

17.5. Isolation of Full-Length cDNA

As discussed previously, the likelihood of obtaining full-length cDNA molecules is increased when all ribonuclease activities have been removed during the first reaction step. When the amount of double-stranded cDNA is sufficient and when its expected size is known, further enrichment for the cDNA of interest can be achieved by running the preparation on a denaturing agarose gel and excising the band of expected molecular weight.

Nevertheless, even the longest cDNA molecules usually lose 15–20 nucleotides at their 5′ end. This phenomenon results from the hydrolysis of the 3′ hairpin of double-stranded cDNA by S1 nuclease. In order to obtain full-length cDNA, Gething et al. (1980) have proposed to use the incomplete cDNA synthesized as described above as a (rather long) primer in the reverse transcriptase reaction of the mRNA of interest.

Today, the synthesis of cDNA molecules from mRNA templates is performed without the S1 nuclease step. The alternative technique involves the combined action of DNA polymerase I and *RNase H* in the second step and a final treatment of the double-stranded cDNA with T4 DNA ligase and dNTPs. It is described in detail in chapter 22.

EXPERIMENTS

17.1. Estimating the Required Amount of Reverse Transcriptase

Prepare the reaction buffer by mixing 2.5 μl of a cold dNTP mix, 3.4 μg of poly(A)$^+$ mRNA (for the example in Fig. 17.4) resuspended in 5 μl of TE and denatured by 2-min incubation at 70 °C, 2 μl of 700 mM β-mercaptoethanol, 10 μl of oligo(dT)$_{12-18}$, 5 μl of 1 M Tris-HCl (pH 8.3), 7 μl of 1 M KCl, 2 μl of 250 mM MgCl$_2$, and 10 μl of ^{32}P-labeled dCTP, for a final volume of 43.5 μl. Split the solution into 4 Eppendorf tubes (10.9 μl per tube). Add 0.5, 1.0, 1.5, and 2 μl of reverse transcriptase and 12 μl of H$_2$O per tube. Incubate the tubes at 42 °C for 2 h and stop the reaction by addition of 0.5 μl of 500 mM EDTA, pH 8.0.

Figure 17.4 shows the percentage of ^{32}P incorporated as a function of the amount of reverse transcriptase added, following TCA precipitation, filtering,

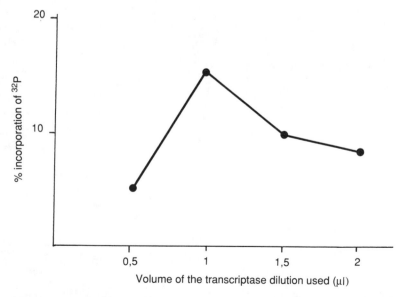

Figure 17.4. *Example of optimization of the volume of reverse transcriptase used in the synthesis of a particular cDNA.*

and scintillation counting. In this particular example, maximum incorporation is obtained when 1 μl of the reverse transcriptase preparation is used.

17.2. Synthesis of the First cDNA Strand

The reaction is carried out in a 500-μl final volume that consists of 50 mM Tris-HCl (pH 8.4); 60 mM NaCl; 10 mM MgCl$_2$; 10 mM DTT; 50 μg/ml oligo(dT)$_{12-18}$; 100 μg/ml actinomycin; 1 mM each of dATP, dGTP, and dTTP; 50 μM of [^{32}P]-dCTP at 65 mCi/mol; 100 μg/ml BSA, 200 μg/ml of poly(A)$^+$ mRNA; and the appropriate amount (in this case 1,250 U/ml) of reverse transcriptase.

Incubate the mixture at 48 °C for 1 h and stop the reaction by adding 20 μl of 10% SDS and 20 μl of 500 mM EDTA. Hydrolyze the mRNA template by adjusting the reaction mixture to a final NaOH concentration of 0.3 M and incubating it for 1 h at 70 °C. Neutralize with acetic acid to a 0.3 M final concentration.

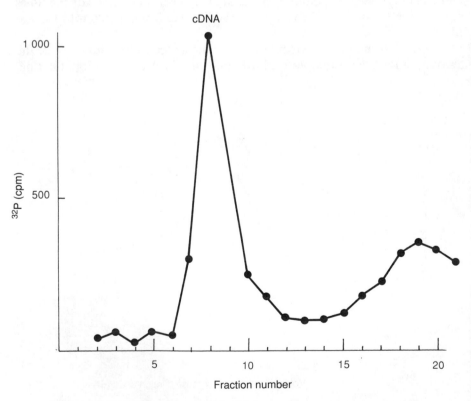

Figure 17.5. *Elution of single-stranded cDNA.*

Remove a small aliquot and precipitate by addition of ice-cold TCA to a final concentration of 8%. Filter the precipitate and count to determine the synthesis yield. Eliminate the unincorporated nucleotides by passing the reaction mixture through a G-200 Sephadex matrix equilibrated and developed with 50 mM Tris-HCl (pH 7.5), 200 mM LiCl, 1 mM EDTA. Collect 500-μl fractions and count small aliquots (Fig. 17.5). Pool the high ^{32}P activity fractions that correspond to the single-stranded cDNA and precipitate with two volumes of ethanol overnight at $-20°C$.

17.3. Synthesis of the Second cDNA Strand

Recover the single-stranded cDNA obtained in experiment 2 by centrifugation at 37,000g for 90 min. Wash the pellet with ethanol and resuspend in a small volume of 500 mM EDTA.

Synthesis of the second DNA strand is performed under the reaction conditions described above except that oligo(dT) and actinomycin are omitted and [^{32}P]-dCTP is replaced by [^{3}H]-dCTP (5 Ci/mmol) at a 90 μM final concentration. As previously, the reaction is stopped with SDS-EDTA and the unincorporated nucleotides are removed by chromatography through a Sephadex G-200 column. Analyze the fractions by counting small aliquots both in the ^{32}P and ^{3}H spectra. Pool the fractions with high ^{32}P and ^{3}H activity that correspond to the double-stranded cDNA and ethanol precipitate overnight at $-20°C$.

Table 17.1 Counts Per Minute Measured In The Different Fractions Collected Following The First Reaction: Single-Stranded cDNA Corresponds To Fractions 4–6

Fraction No.	cpm
1	13
2	14
3	18
4	2,614
5	5,036
6	1,312
7	23,931
8	146,579
9	259,006
10	187,614
11	89,639
12	18,170
13	2,604
14	315
19	117

17.4. S1 Nuclease Digestion

Recover the precipitated DNA by centrifugation as described above, wash and dry the pellet, and resuspend in a small amount of 500 mM EDTA. Adjust to 40 mM sodium acetate (pH 4.5), 266 mM NaCl, and 5 mM $ZnSO_4$ for a final reaction volume of 500 μl. Add the appropriate quantity of S1 nuclease (in the example, 6,000 U/ml). Incubate for 1 h at 25 °C and stop the reaction by adding 100 mM Tris-HCl, pH 8.4. Remove a small aliquot and verify the efficiency of the digestion by TCA precipitation (8% final concentration) followed by filtration and counting. Purify the cDNA by phenol–chloroform extraction and pre-

Figure 17.6. *Autoradiogram of a double-stranded cDNA preparation on a denaturing agarose gel before (a) and after (b) treatment with S1 nuclease.*

cipitate by adding NaCl to a 200 mM final concentration, and two volumes ethanol, and incubate overnight at $-20\,°C$.

PROBLEMS

17.1 Yield of Single-Stranded cDNA

Table 17.1 shows the amount of counts present in each fraction following synthesis of single-stranded cDNA and fractionation on a Sephadex G-200 column. Knowing that 10 μg of mRNA was used in the reaction and that the final concentration of labeled dCTP ($M_r = 467$) was 0.1 mM for a reaction volume of 1 ml, calculate the yield of single-stranded cDNA (which elutes in fractions 4, 5, and 6) obtained from the mRNA template.

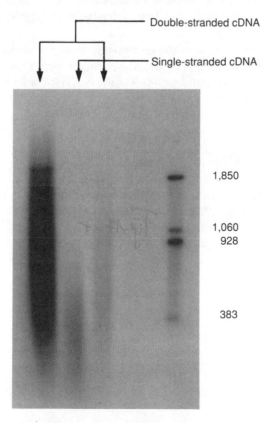

Figure 17.7. *Autoradiogram of a ^{32}P-labeled preparation of cDNA synthesized from human liver poly(A)$^+$ mRNA.*

17.2 Yield of Double-Stranded cDNA

In spite of the precautions taken during synthesis, the final amount of double-stranded cDNA generally represents only a few percentage points of the initial mass of mRNA used in the first reaction. Knowing that 20–100 ng of cDNA is required for cloning purposes, what amount of poly(A)$^+$ mRNA should one start with?

17.3 Verifying the cDNA Digestion by S1 Nuclease

Figure 17.6 shows the autoradiogram of a double-stranded cDNA preparation before and after treatment with S1 nuclease. What do you conclude from this gel?

17.4 Heterogeneous Nature of Double-Stranded cDNA Prepared From Total Liver Poly(A)$^+$ mRNA

Figure 17.7 shows the results of the analysis of double-stranded cDNA synthesized from human liver poly(A)$^+$ mRNA. Samples were resolved on a 1% agarose gel under denaturing conditions (30 mM NaOH, 2 mM EDTA). What do you conclude after inspecting the figure? Within the smear, discrete bands can be distinguished; what do such bands correspond to?

17.5 Incomplete Double-Stranded cDNA

When a preparation of double-stranded cDNA molecules obtained by the action of reverse transcriptase is analyzed by electrophoresis, only a few molecules correspond to the full length of the message; most cDNAs are shorter. How do you explain the synthesis of these short molecules?

Part V
Cloning Techniques

18

Ligation

Ligation is a process in which two cleaved DNA fragments (single- or double-stranded) are joined to each other through the enzymatic action of a *DNA poly-nucleotide kinase*. This enzyme catalyzes the formation of a *phosphodiester bond* between the 3′-hydroxyl and 5′-phosphate groups of the nucleotides located on separate DNA fragments in order to yield a continuous molecule. DNA ligases play an essential role in vivo during DNA replication, repair, and recombination. In vitro, DNA ligases are important for all genetic engineering applications since they permit the joining of blunt or staggered-ended DNA molecules together. A typical application consists of using a ligase to secure a passenger DNA molecule in a cloning vector.

18.1. Ligases

Two ligases are commonly used in genetic engineering:

- The *E. coli* DNA ligase (M_r = 74,000) is capable of covalently joining the two extremities of the same DNA strand. The enzyme requires NAD^+ as a cofactor and its activity is strongly stimulated in the presence of NH^{4+} ions.
- The most widely used enzyme is *T4 DNA ligase*. It is synthesized by gene *30* from bacteriophage T4, which encodes its own ligase. T4 DNA ligase (M_r =

69,000) is fully dependent upon the presence of ATP and, in contrast to the *E. coli* ligase, can join either blunt or staggered-ended DNA fragments.

In both cases, NAD^+ or ATP is involved in the formation of an enzyme–adenylate complex. Before a phosphodiester bond can be created, the complex binds to DNA at the cleavage site, which results in exposing the 3'-OH and 5'-P groups. The adenylyl residue is then transferred to the DNA to form a DNA–P–P–adenylate complex. Next, a phosphodiester bond is formed with the 3'-OH group and, depending on the enzyme, NMN or AMP is released.

18.2. DNA Molecules Involved in Ligations

Cloning operations involve two different types of DNA molecules (Fig. 18.1). The first is the DNA fragment containing the gene of interest (also called *passenger DNA*), and the second is the cloning vector DNA. If both the passenger DNA and the vector are digested by a restriction enzyme that generates staggered ends (e.g., *Eco*RI in the case of Fig. 18.1) and mixed, the two molecules can associate by annealing their single-stranded, complementary ends. The junction regions, where the DNA backbone is broken, are recognized by ligases and a phosphodiester bond is formed, covalently linking the DNA strands. A similar result is obtained when blunt-ended restriction enzymes are used, except that no annealing takes place and that the efficiency of the reaction is lower.

Several types of molecules can be generated during ligation reactions. Obviously, the form of interest is a recombinant molecule that consists of one extraneous DNA fragment ligated within one copy of the vector DNA. However, other scenarios may also occur. For instance, the vector can reanneal to itself and be regenerated by ligation. In addition, polymeric molecules consisting of multiple copies of the passenger DNA or vector DNA may be formed by self-annealing. These long chains may next circularize and become ligated to themselves or to each other. In this fashion, recombinant DNA molecules containing multiple passenger DNAs ligated within one or several copies of vector DNA can be formed. The different techniques used to minimize these undesirable ligation products are discussed below.

18.3. In Vivo vs. In Vitro Ligation

The first ligation experiments were accomplished in vivo. It soon became obvious that the process was cumbersome and complicated since it required the use of staggered-ended DNA molecules able to anneal to each other under the physiological conditions encountered in the cell. In addition, the in vivo ligation reactions, performed by the host-cell DNA-repair system, only yielded a small number of recombinants, which often contained small deletions at the

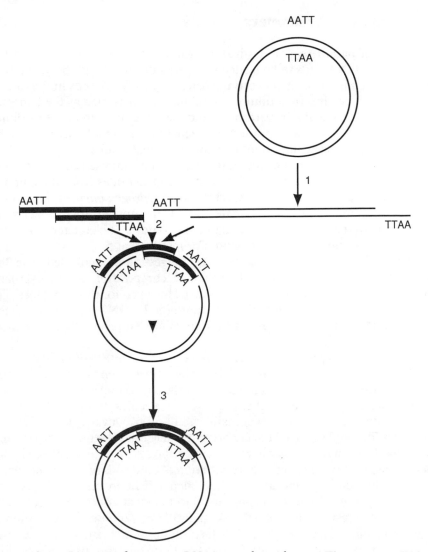

Figure 18.1. *Ligation of passenger DNA into a plasmid vector.* The passenger DNA is mixed with vector DNA linearized with a compatible restriction enzyme (1). The extremities of the two DNA species hybridize (2) and are covalently joined by T4 DNA ligase (3).

site of ligation. As a result, ligations are today almost always performed with purified enzymes. The in vitro approach presents several advantages. First, the degradation of DNA fragments is eliminated since the commercially available ligase preparations are devoid of DNase activity. Second, the ligation conditions can be optimized in order to maximize the number of recombinant DNA molecules.

18.4. Ligation Parameters

A number of parameters are critical in ligation reactions. They include the nature of the DNA ends to be ligated (staggered or blunt) and the strength of the base–base interactions between the heterologous DNA ends in the case of staggered-end termini. In addition, several other factors such as the temperature at which the reaction is performed and the ionic strength of the medium, as well as the relative concentrations and length of the DNA fragments to be ligated, are particularly important to achieve optimal results.

When restriction enzymes generating sticky ends are used to cut both the vector and the DNA fragment of interest, the nucleotides from the complementary overhangs are able to interact through *hydrogen bonding*. Since only a few nucleotides are involved, a small increase in temperature is sufficient to break the hydrogen bonds. The strength of the binding is characterized by the *melting temperature* (Tm) of the bond. For instance, at 5 °C, half of the staggered 5′-AATT-3′ ends generated by *Eco*RI digestion are annealed. The Tm value varies with the length and nucleotide composition of the overhang (G−C bonds are stronger than A−T bonds). However, for most enzymes, Tm is about 15 °C. Since the optimal temperature for T4 DNA ligase is 37 °C, the reaction is generally carried out at an intermediate temperature (varying from 12.5 to 16 °C).

Ligases are able to produce both *intramolecular* and *intermolecular* covalent bonds. Intramolecular bonds are obtained when the two ends of the same molecule are ligated to generate a circular monomer. Intermolecular bonds result from the ligation of the ends of two different molecules; this operation yields linear multimers that can be eventually circularized. Dugaiczyk et al. (1975) have identified and modeled the factors that determine whether intra- or intermolecular ligation reactions will predominate. Two parameters are used in this model: j represents the effective concentration of both ends of a given molecule and i the total end concentration. While i depends upon the total DNA concentration, j is only inversely proportional to the length of the molecule. The proximity of the two ends of a DNA molecule of random conformation is related to its molecular weight and its three-dimensional structure. The latter parameter depends itself on the *ionic strength* of the reaction medium. For instance, if an excess of Na^+ cations is used, the DNA double helix becomes more rigid and the effective end concentration decreases. Therefore, the standard buffer used in ligation reactions is designed to minimize ionic effects.

The j value (in ends per ml) can be calculated for any DNA molecule by using the formula:

$$j = j\lambda \cdot \left(\frac{mw_\lambda}{mw} \right)^{3/2} \tag{1}$$

where j_λ and mw_λ are the effective end concentration and molecular weight for linearized λ DNA ($j_\lambda = 3.6 \times 10^{11}$ ends/ml, $mw_\lambda = 30.8 \times 10^6$ Da) and j and mw are the effective end concentration and molecular weight of the DNA molecule of interest.

In the case of DNA molecules with identical staggered ends, the total end concentration is calculated from (2):

$$i = 2 \cdot N_0 \cdot M \cdot 10^{-3} \tag{2}$$

where $N_0 = 6.02 \times 10^{23}$ is Avogadro's number, and M is the molar concentration of the DNA molecules.

The j/i ratio can therefore be estimated by replacing j_λ, pm_λ, and N_0 with their appropriate values and using the fact that M is related to the DNA concentration in the sample (noted [DNA] and expressed in g/L) by the equation:

$$[DNA] = \frac{51.1}{(mw)^{1/2}}$$

Hence:

$$\frac{j}{i} = \frac{51.1}{[DNA] \cdot (mw)^{1/2}} \tag{3}$$

Theoretically, when $j/i < 1$, intermolecular reactions are favored, while when $j/i > 1$ intramolecular reactions predominate. However, experimental studies have demonstrated that intramolecular reactions only occur when $j/i > 2$ or 3. In addition, the j/i ratio varies as ligation proceeds since the total end concentration decreases while j remains constant for the fragment analyzed. Thus, there is a progressive tendency to circularization of the molecules with time.

It is possible to manipulate the previous formula to obtain mw:

$$mw = \left(\frac{51.1}{j/i \cdot [DNA]} \right)^2 \tag{4}$$

Figure 18.2 shows how mw varies with the DNA concentration for different j/i ratios. By using this figure, the structure of the ligation products of a DNA fragment of known molecular weight can be estimated as a function of its concentration. As the j/i ratio increases, DNA circularization is favored, while as j/i decreases linear concatemers are more easily formed. The j/i coefficients shown in Fig. 18.2 have been calculated for the ligation of DNA fragments of identical molecular weight, which is rarely the case experimentally. Nevertheless, one can get a good idea of the way the reaction will proceed by using the average molecular weights of the fragments to be ligated. For instance, if a 2×10^6 Da fragment is to be ligated with a 1.6×10^6 Da vector, the j/i ratio equals 1 when the DNA concentration is 55 μg/ml. Under these conditions, the formation of concatemers is favored. (Some of these will be homologous concatemers corresponding to the ligation of either DNA fragment in the reaction mixture to the other.) When the 2×10^6 Da fragment is present at a 10-fold molar excess relative to the vector DNA concentration, vector dimerization will occur at a frequency of one out of 10 recombinants. Thus, as a general rule, an excess of the large fragment in a ligation reaction favors the formation of multimers of this DNA. However, it is also beneficial to the formation of

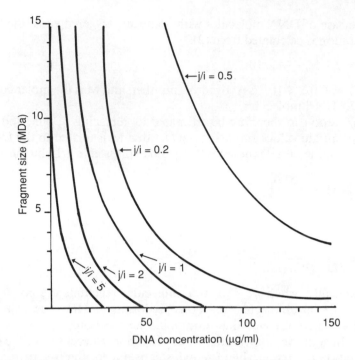

Figure 18.2. *Relationship between molecular weight and DNA concentration for five values of the j/i ratio* (source, BRL).

recombinant molecules while minimizing the generation of multimers of the small fragment.

18.5. Increasing the Cloning Efficiency by the Simultaneous Use of Two Restriction Enzymes

One of the major inconveniences encountered during the cloning of foreign DNA fragments into vectors resides in the fact that the vector may religate on itself without incorporating the DNA sequence of interest. Since the religated vector is smaller than the recombinant plasmid, it can transform competent cells at a higher efficiency thereby lowering the yield of recombinants.

A typical approach to prevent this problem is to digest vector and exogenous DNA each with two different restriction enzymes in order to generate two incompatible ends on each linearized fragment. If one uses this technique, no intramolecular ligation can take place and a circularized plasmid is obtained only when each end of the insert anneals with the corresponding compatible ends of the vector. It is also possible to digest vector and insert DNA with restriction enzymes that have different recognition sequences but generate the same overhang. For instance, the cloning vector can be cleaved with *Bam*HI

(G ↓ GATCC) and the foreign DNA with *Bcl*I (T ↓ GATCA). Religation of the fragments will generate recircularized vector, circularized foreign DNA, and recombinants. However, the recombinants will contain a hybrid junction (GGATCA) that is no longer recognized by either enzyme. Thus, treatment of the ligation mix with *Bam*HI will only linearize the recircularized vector molecules. Since linear DNA transforms competent cells at very low efficiency, most colonies will harbor the recombinant plasmid following transformation. Another advantage of the latter method is that the insert can be readily excised from the recombinant plasmid by digestion with *Sau*3AI (↓ GATC). Finally, both methods have the advantage of yielding recombinant plasmids containing a passenger DNA in a given orientation *(directional cloning)*.

18.6. Treatment of Linearized DNA With Alkaline Phosphatase

Another method to eliminate the background of recircularized plasmids consists of treating linearized vectors with the enzyme *alkaline phosphatase* (Appendix 19), which removes the 5'-phosphate groups exposed by restriction enzyme digestion. Following this treatment, the vector becomes unable to recircularize or form dimers (Ulrich et al., 1977) unless a 5'-phosphate is provided by a foreign DNA fragment. The resulting recombinant plasmid still contains a nick at each extremity, but it is filled by the repair system of the host cell following transformation.

18.7. Blunt-End Ligations

Although T4 DNA ligase catalyzes the ligation of extremities generated by digestion with either stagger-ended or blunt-end enzymes, the latter process is more complex and carried out with a lower efficiency. Typically, blunt-end ligations occur 100-fold slower compared to staggered-end ligations. This presumably results from the lower probability for two unrelated extremities to meet. In fact, it has been proposed that blunt-end ligations require two different types of ligases: the first would bring the ends together; the second would catalyze the formation of a phosphodiester bond. In general, 10- to 30-fold-higher ligase concentrations are required to ligate blunt ends with the same efficiency as staggered ends. Although the juxtaposition of blunt ends is facilitated by increasing the DNA end concentration, most blunt-end ligations are carried out with end concentrations in the μmolar range for a j/i ratio smaller than 0.5. In most cases, the optimal ligation temperature is close to the Tm value of the smaller DNA fragment to be ligated. Finally, blunt-end ligations are reversibly inhibited in the presence of an excess of ATP. (Although a 2.5 mM concentration of ATP is optimal for both types of ligations, it also selectively inhibits the ligation between blunt-ended DNA molecules.)

EXPERIMENTS

18.1. Typical Ligation Protocol

In a fresh Eppendorf tube, mix 1 μg of linearized vector, 1 μg of linearized insert, 5 μl of 10× ligation buffer (250 mM Tris-HCl, pH 7.6; 100 mM $MgCl_2$; 10 mM DTT; 4 mM DTT; 0.1 unit T4 DNA ligase) and adjust the volume to 50 μl with ddH_2O. Incubate at 12 °C for at least 12 h.

18.2. Blunt-End Ligations

Mix linearized vector and linearized insert at a 1:3 ratio for a final DNA amount of 200 ng. Add 2 μl of 10× ligation buffer and adjust the volume to 20 μl with ddH_2O. Incubate at 20 °C for 12 h.

18.3. Vector Dephosphorylation

To the digested vector resuspended in a final volume of 10 μl, add 5 μl of 10× alkaline phosphatase buffer (500 mM Tris-HCl, pH 8.0; 500 mM NaCl; 1 unit alkaline phosphatase). Adjust the volume to 100 μl with ddH_2O and incubate at 65 °C for 1 h.

PROBLEMS

18.1 Schematic Representation of Ligation Reactions

Using a figure, show the mode of action of T4 DNA ligase in the following cases:

a. Ligation of a nick in one of the DNA strands of a double-stranded molecule
b. Staggered-end ligation (assume 4 bp overhangs)
c. Blunt-end ligation

18.2 Efficiency of Ligation and Nature of Overhangs

Explain why, although both overhangs contain the same number of nucleotides, DNA molecules digested with PstI are religated much faster than those cleaved with EcoRI.

18.3 Restriction Enzymes and Ligation

What do restriction enzymes *Sal*I and *Xho*I have in common? How can they be used in cloning?

18.4 Determination of End Concentrations

Calculate the concentration of extremities in pmol/µg of vector DNA to clone an insert at the *Pst*I site of pBR322 or λ. Use the empirical formulae:

$$[\text{pBR322}] = \frac{2N \times 10^6}{mw}$$

and

$$[\lambda] = \frac{2(N + 1) \times 10^6}{mw}$$

where N is the number of restriction sites and mw the molecular weight.

18.5 Calculating $j\lambda$

How do you calculate the effective end concentration in the vicinity of the same DNA molecule (j value) for bacteriophage λ?

18.6 Influence of the DNA Concentration on the j/i Ratio

Using Fig. 18.2, give the values of the j/i ratio when a 4×10^6 Da DNA fragment is ligated at 10 µg/ml or 50 µg/ml concentrations. What type of ligations will predominate under such conditions?

Cloning by Addition of Complementary Homopolymeric Sequences

A useful method for the cloning of cDNA fragments consists of artificially elongating their extremities with complementary *homopolymeric sequences* (Jackson et al., 1972). Figure 19.1 depicts a schematic representation of a technique that has been widely used in the past two decades. Briefly, the passenger cDNA, elongated with poly(C), is cloned into the *Pst*I site of pBR322, whose ends have themselves been elongated with poly(G) (Villa-Komarof et al., 1978). This cloning method does not require a ligation step and facilitates the identification of recombinant plasmids since they display an ampicillin-sensitive, tetracycline-resistant phenotype. It also regenerates the *Pst*I site, which greatly simplifies the excision of the cDNA of interest after sizable quantities of the recombinant plasmid have been isolated.

For the reaction shown in Fig. 19.1, homopolymeric sequences are synthesized by the enzyme *deoxynucleotidyl terminal transferase* (Appendix 20), which sequentially adds deoxynucleotides at the 3′ end of linear DNA molecules. The reaction is performed in a cacodylate buffer and does not require a template. The enzyme is maximally active on 3′ overhangs but only weakly active on 3′ receding ends. Blunt-ended DNA molecules are normally a poor substrate for terminal transferase; however, such ends may still be extended if cobalt ions are substituted for magnesium ions in the reaction buffer (Brutlag et al., 1977). This peculiar effect is explained by the fact that cobalt ions relax the helical structure of DNA.

Since terminal transferase is not a specific enzyme, any of the four dNTPs can be used to synthesize 3′ extensions composed of only one type of residue *(homopolymeric tracts)*. A typical procedure consists of synthesizing poly(dC)

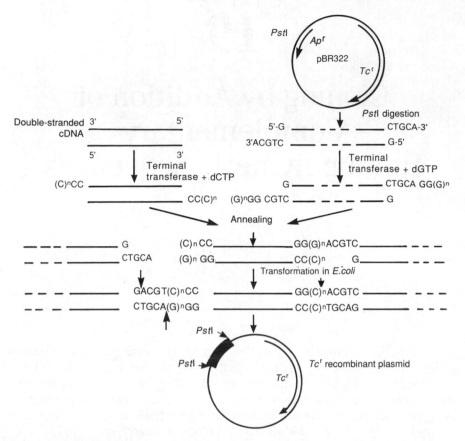

Figure 19.1. *Cloning in the* Pst*I site of pBR322 using homopolymeric tracts.* The following steps are illustrated: addition of poly(C) tracts at the 3′ ends of the double-stranded cDNA using terminal transferase; digestion of pBR322 with *Pst*I and elongation of its 3′ ends with poly(G) tracts; hybridization of the ends of the two DNA species through G−C base pairing; transformation of the resulting plasmid into *E. coli*. (The host-repair system will fill out the gaps in the hybridized domains and thus regenerate the *Pst*I sites.) The resulting recombinant plasmid is *Ap^s* and *Tc^r*.

tails at the extremities of the DNA fragment to be cloned while generating poly(dG) extensions at the ends of the linearized vector DNA. The complementary tracts readily hybridize to form a recombinant molecule after treatment with T4 DNA ligase. The length of the newly synthesized tails can be determined by using radiolabeled nucleotides. Typically, 100 dA/dT or 20 dC/dG residues must be added to the DNA fragments to achieve efficient annealing and transformation (Peacock et al., 1981).

When the passenger DNA is to be cloned into the *Pst*I site of plasmid pBR322, extension of the cDNA with poly(dC) homopolymeric tails is relatively straightforward. It is, however, critical to accurately measure the concentration of cDNA molecules used in the reaction. On the other hand, great care must be taken in the preparation of the plasmid DNA. The vector must be free of any contaminating chromosomal DNA to avoid cloning artifacts. Therefore, the cloning vector is usually purified by two successive ultracentrifugation rounds on CsCl gradients. In addition, *Pst*I digestion must be carried out to completion in order to minimize the background. Finally, excess concentrations of restriction enzyme must be avoided since contaminating endonucleases may hydrolyze the 3' overhangs and thus prevent the reconstitution of the *Pst*I site. When the vector linearized with *Pst*I and elongated with poly(dG) is mixed with a cDNA elongated with poly(dC), a ligation step is not necessary. Figure 19.1 shows that upon annealing, two short, single-stranded sequences are formed; however, these "holes" are filled by the *rec*A-dependent repair system of the host upon transformation.

Depending on the type of cloning procedure used, different methods can be selected to isolate the insert of interest (which is now available in large quantities). If cloning is carried out with dG/dC extensions at the *Pst*I site of pBR322, the passenger DNA can be recovered by *Pst*I digestion and purified on a low-melting-point agarose gel. If dA/dT extensions are used for cloning, it is possible to obtain the fragment of interest by subjecting the recombinant plasmid DNA to gentle denaturation in order to break the A − T bonds of the homopolymeric tracts. The denatured single-stranded domains are then hydrolyzed by S1 nuclease treatment. Other protocols are also useful in the case of dA/dT extensions. The first consists of cutting pBR322 at its *Cla*I site, to generate poly(dA) and poly(dT) tails on the cDNA and the vector, respectively. Since the *Cla*I site is flanked by *Eco*RI and *Hin*dIII sites, digestion of the recombinant plasmid by these enzymes will yield the desired fragment. Furthermore, pBR322 can be linearized with *Hin*dIII, the 3' receding ends can be filled with Klenow, and poly(dT) tails can be synthesized using terminal transferase. When the modified vector is annealed to a cDNA displaying poly(dA) tails, the recombinant plasmid will contain regenerated *Hin*dIII sites that can be used to isolate the passenger DNA.

An elegant technique for the isolation of full length cDNA molecules was described by Land et al. (1981) and is illustrated in Fig. 19.2. The first cDNA strand synthesized by reverse transcriptase is elongated with poly(dC) using terminal transferase and recovered following mRNA hydrolysis. The second strand of the cDNA molecule is then synthesized by reverse transcriptase, using oligo(dG) as a primer. This technique is more efficient at yielding full-length double-stranded cDNA molecules than the classic approach employing S1 nuclease.

A similar approach can be used in combination with the *polymerase chain reaction* (PCR, Appendix 21) to generate large quantities of the cDNA of interest and thus facilitate its cloning. For example, a tract of poly(dG) can be synthesized at the 3' end of the original cDNA obtained from the mRNA template. Since the cDNA already contains a poly(dT) tract at its 5' terminus, amplification can be easily performed by using oligo(dT) and oligo(dC) as primers, as described in Appendix 21. This approach is particularly useful for generating cDNA libraries (see Chapter 22) from a mRNA pool isolated from 1 or 2 mam-

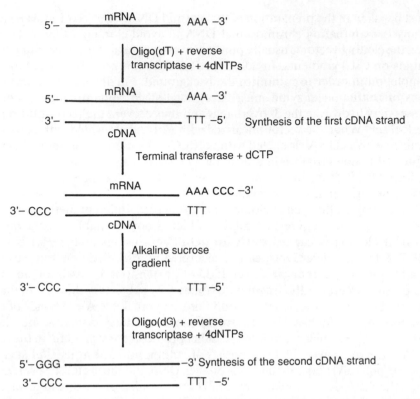

Figure 19.2. *cDNA elongation.* The 3′ end of the first cDNA strand is elongated with oligo(C). An oligo(G) oligonucleotide can anneal to this region and serve as a primer for the synthesis of the second cDNA strand.

malian cells. Its major inconvenience lies in the fact that the amplified sequences can accumulate a large number of errors.

EXPERIMENTS

19.1. Synthesis of Poly(dC) Homopolymeric Tracts at the 3′ End of Double-Stranded cDNA Molecules (From Roychoudhury et al., 1976)

Recover the double-stranded cDNA of interest by centrifugation and wash the pellet twice with ethanol. Resuspend the dried pellet in a small volume of 0.1 mM EDTA. Perform the elongation reaction with 80 μg/ml of double-stranded cDNA in a 25-μl final volume of elongation buffer (30 mM Tris-HCl, pH 7.6; 140 mM sodium cacodylate; 0.1 mM DTT; 3.5 mM CoCl$_2$; 0.08 mM [^{32}P]-

dCTP at 2 Ci/mmol). Warm the mixture to 37 °C, add 1,000 U/ml of deoxynucleotide terminal transferase, and incubate at this temperature. After 10–15 min, verify the progress of the reaction by precipitating a small aliquot in 3 ml of ice-cold TCA and counting. When the desired number of nucleotide per extremity is reached, stop the reaction by addition of EDTA to a 10 mM final concentration.

Remove the contaminating enzymes by phenol–chloroform extraction and apply the aqueous phase to a Sephadex G50 column to remove unincorporated [^{32}P]-dCTP. Pool the fractions corresponding to the first radioactivity peak and collect the elongated cDNA by ethanol precipitation.

19.2. Preparing pBR322 for the Cloning Step

Linearize 100 μg/ml of pBR322 with 5 units of *Pst*I per μg of DNA in 6 mM Tris-Hcl (pH7.5), 50 mM NaCl, 6 mM MgCl$_2$, 6 mM β-mercaptoethanol, 100 μg/ml BSA. After 60 min at 37 °C, add another 5 units of enzyme per μg of DNA and continue the incubation for an additional hour. Stop with EDTA at a final concentration of 12 mM and phenol–chloroform extract the DNA.

Precipitate the linearized DNA with ethanol and recover the pellet by centrifugation. After an ethanol wash step, perform the elongation reaction using 80 μg/ml of DNA and 1,000 units of terminal transferase per ml of buffer (30 mM Tris, pH 7.6; 140 mM sodium cacodylate; 0.1 mM DTT; 1 mM CoCl$_2$; 0.06 mM [^{3}H]-dGTP at 5 Ci/mmol). Following the extent of elongation by TCA-precipitating small aliquots and counting. It takes 2–3 min incubation at 37 °C to add 15–20 nucleotides to the extremities of linearized pBR322.

19.3. Hybridization and Formation of the Recombinant Plasmid

Mix 10–15 ng of poly(dC)-elongated cDNA to 100 ng of poly(dG)-elongated pBR322 in a 25-μl final volume of 10 mM Tris-HCl (pH 7.5), 100 mM NaCl, 1 mM EDTA. Transfer the tube to a 65 °C water bath and incubate for 1 h at this temperature. Transfer the tube to a 46 °C water bath 1 h and to a 37 °C for 1 h. Allow to slowly cool to room temperature.

PROBLEMS

19.1 Number of dNTP Residues Synthesized per 3′ End

How do you calculate the number of dNTP residues added by terminal transferase per 3′ terminus of DNA?

19.2 Cloning at the *Hin*dIII Site of pBR322 Using dA/dT Tracts

Using an annotated figure similar to Fig. 19.1, show the steps involved in the cloning of a double-stranded cDNA at the *Hin*dIII site of pBR322 using dA/dT extensions.

19.3 Number of 3′ Ends Available for Generating Homopolymeric Tracts

Plasmic pAT (2.5×10^6 Da) contains a unique *Pst*I site. Calculate the weight (in μg) of 1 pmole of *Pst*I ends when the plasmid is linearized with this enzyme. For every pmole of 3′ ends, terminal transferase generates an average elongation of 25 nucleotides with 25 pmol of dC. Given that the specific activity of [^{32}P]-dCTP is 1,000 Ci/mmol, how many μCi of [^{32}P]-dCTP should be added to the reaction? Answer the same question if [^{3}H]-dCTP at a specific activity of 17 Ci/mmol is substituted to [^{32}P]-dCTP.

20

Cloning by Addition of Synthetic Sequences

Although homopolymeric tailing has been widely and successfully used for cDNA cloning, it was superceded when reliable sources of *synthetic linkers* and *adaptors* became available. T4 DNA ligase is able to ligate double-stranded as well as blunt-ended DNA molecules. This property has been exploited to join synthetic linker molecules designed to contain the recognition sequences for different restriction enzymes to blunt-ended cDNA molecules. Following ligation to the cDNA and digestion by the appropriate restriction enzyme(s), the resulting *chimeric* fragment can be readily cloned into the appropriate site(s) of a vector. A different type of synthetic molecule used to achieve the same result, but without a requirement for restriction-enzyme digestion, is the double-stranded *adaptor*. Before synthetic molecules may be ligated to the cDNA, it must be blunted.

20.1. Generating Blunt Ends From Staggered Ends

Two strategies can be used to obtain blunt ends from cohesive (staggered) ends. One consists of hydrolyzing the producing ends by nucleases; the other consists of extending the receding ends with a polymerase.

The first approach involves the enzyme S1 nuclease, which, in high-ionic-strength buffers (>200 mM NaCl) and at pH 4.0, specifically hydrolyzes single-stranded DNA. The reaction must be carefully controlled since, at high concentration, S1 nuclease may also hydrolyze denatured regions in double-stranded cDNA molecules. Therefore, the reaction is usually carried out at low temperatures (about 25 °C).

The second strategy employs *E. coli* DNA polymerase I, its Klenow fragment, or T4 DNA polymerase. When a 3′ receding end is obtained by digesting DNA with enzymes (such as *Eco*RI or *Hin*dIII), any of the above modifying enzymes is capable of filling in the gaps with nucleotides.

$$\ldots G - 3' \qquad \underrightarrow{\text{DNA polymerase}} \qquad \ldots GGATC - 3'$$
$$\ldots CCTAG - 5' \qquad + \text{dNTPs} \qquad \ldots CCTAG - 5'$$

In the case of a 3′ protruding end (e.g., as obtained with *Pst*I or *Kpn*I), the extension can be removed using the 3′ → 5′ exonuclease activity of either *E. coli* or T4 DNA polymerase. Since the exonuclease activity can continue hydrolysis in the 3′ → 5′ direction, the enzymes are used in the presence of an excess of dNTPs, which prevents unwanted degradation by favoring polymerization as soon as a template becomes available.

$$E.\ coli \text{ or } T4$$
$$\ldots CTGCA - 3' \ \underrightarrow{\text{DNA polymerase}} \ \ldots C - 3'$$
$$\ldots G - 5' \qquad + \text{dNTPs} \qquad \ldots G - 5'$$

In some cases, the *Bal*I nuclease, which possesses two exonuclease activities (3′ → 5′ and 5′ → 3′), is also used to generate blunt-ended DNA molecules. However, the extent of hydrolysis is much more difficult to control with this enzyme.

20.2. Synthetic Linkers

Linker molecules are generally octo- or decameric single-stranded oligodeoxynucleotides that are artificially synthesized as palindromic sequences. They can be annealed by heating followed by slow cooling and thus generate a blunt-ended double-stranded DNA fragment containing the recognition site for a restriction enzyme (Bahl et al., 1976). An example of an *Eco*RI linker is:

Linker molecules obtained in this fashion (Appendix 22) can be ligated to a blunt-ended cDNA with T4 DNA ligase. Upon digestion by the appropriate restriction enzyme (*Eco*RI in the example of Fig. 20.1), the modified cDNA can be readily cloned in a vector linearized by the same enzyme (Ferreti and Sgaramella, 1981). Figure 20.2 shows a cloning protocol in which *Pst*I linkers and *Pst*I digestion are used to clone the cDNA of interest in the *Pst*I site of a vector.

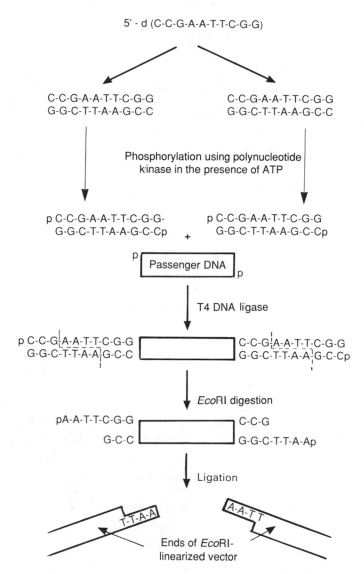

Figure 20.1. *Cloning of a passenger cDNA in the EcoRI site of a plasmid after ligation of synthetic EcoRI linker molecules. Eco*RI oligonucleotide linkers are mixed and allowed to hybridize. Their 5′ ends are phosphorylated by polynucleotide kinase in the presence of ATP. The synthetic linkers are linked to the blunt, double-stranded cDNA with T4 DNA ligase, and *Eco*RI overhangs are generated by hydrolyzing the construction with this enzyme. The modified cDNA is then cloned into a plasmid vector linearized with *Eco*RI.

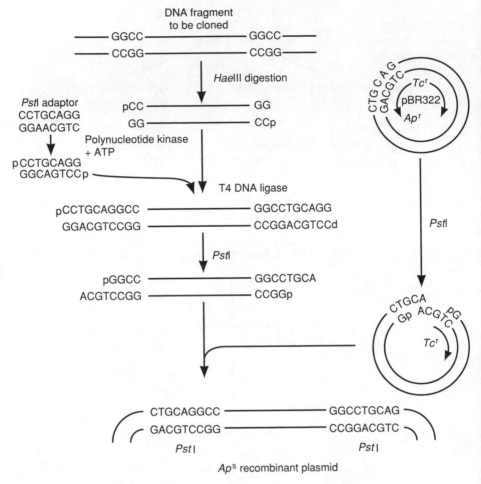

Figure 20.2. *Use of PstI linker molecules for cloning in the PstI site of pBR322.*

Linkers are commercially available as 5'-hydroxylated oligodeoxynucleotides. Consequently, it is necessary to phosphorylate their 5' ends with the enzyme *T4 polynucleotide kinase* before they can be ligated to the cDNA molecules.

The ligation reaction, which joins the linkers to the cDNA, is generally carried out in the presence of an excess of the double-stranded oligonucleotides. Since these molecules can compete with the modified cDNA in the following reactions, it is necessary to remove them from the preparation. A number of approaches have been devised to address this problem. They include gel filtration, separation by electrophoresis, and chromatography on ultrogel. A useful method is specific precipitation with *spermine* (Hoopes and McClure, 1981).

Another consequence of using an excess of linker to cDNA during the first ligation is that the cDNA can be bound to polymeric repeats of the linker. However, this is generally not a major problem since digestion with the appropriate
restriction enzyme will only generate cDNA molecules flanked by one repeat
of the linker on each side.

At this stage, a serious problem is the unproductive recircularization of the
modified cDNA on itself by annealing and ligation of the complementary linkers. To prevent this reaction, the ligation step destined to yield recombinant
molecules is usually performed with a molar excess of linearized vector to modified cDNA. (Under such conditions the vector will also have a tendency to
recircularize; this is usually prevented by dephosphorylating its extremities
with alkaline phosphatase.) Alternatively, two different molecules may be
ligated to the cDNA, which becomes unable to anneal to itself. By cleaving the
vector with the appropriate pair of enzymes (Fig. 20.3), the insert can be cloned
in the desired orientation.

20.3. Adaptors

When the cDNA of interest contains one or several restriction sites corresponding to those encoded by linkers, adaptors may be used. Adaptors are presyn-
thesized as short, double-stranded DNA molecules which have one blunt and
one cohesive end (Bahl et al., 1978). Figure 20.4 shows, as an example, the
modality of use of a *Bam*HI adaptor for cDNA cloning.

EXPERIMENTS

20.1. Phosphorylation of Synthetic Linkers

To 15 μl of 5× kinase buffer (250 mM Tris-HCl, pH. 8.0; 50 mM MgCl$_2$; 75
mM DTT; 50 μM ATP), add 50 pmol of linker and 100 units of T4 polynucleotide kinase and bring the volume to 75 μl with ddH$_2$O. Incubate at 37 °C for
30 min.

20.2. Ligation of a Phosphorylated Linker to a Blunt-Ended cDNA Molecule

In a fresh Eppendorf tube, mix 50 pmol of phosphorylated linker with 1 pmol
of blunt-ended cDNA. Add 2 μl of 10× ligase buffer and 0.2 units of T4 DNA
ligase and complete with 20 μl of ddH$_2$O. Incubate for 12 h at 4 °C.

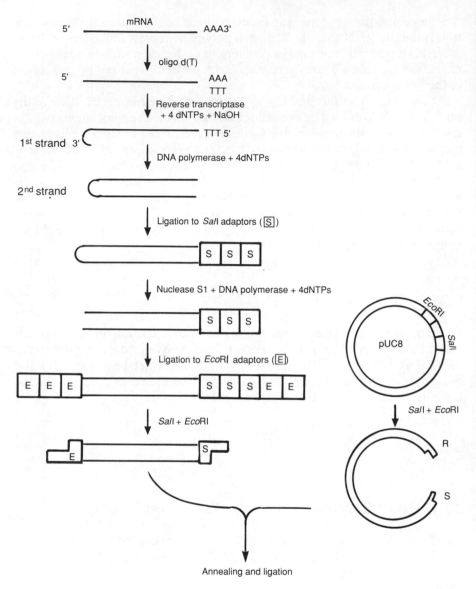

Figure 20.3. *Cloning of a cDNA in pUC8 using two different linker molecules (SalI and EcoRI).*

PROBLEMS

20.1 Regenerating a Restriction Site with Polymerase

Explain how polymerases can regenerate an *Eco*RI site from two termini cleaved with *Eco*RI and *Bam*HI.

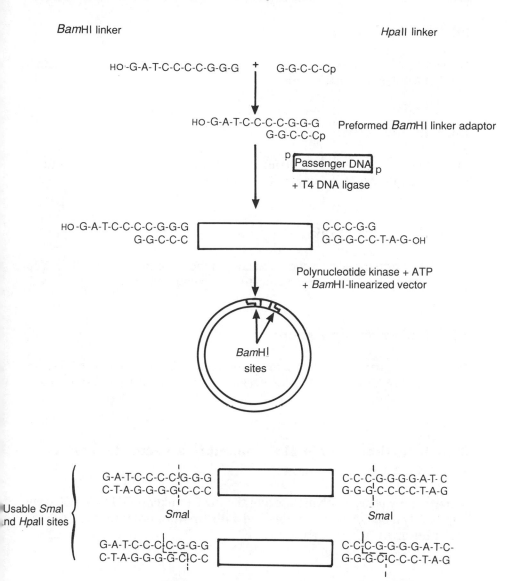

Figure 20.4. *Use of BamHI and HpaII linker molecules to generate a BamHI adaptor.* Since the *Hpa*II linker contains a phosphorylated 5′ end, it can be ligated to the blunt, double-stranded cDNA. Once ligated, the adaptor displays hydroxylated ends, which prevents the modified cDNA from annealing to itself, but cloning at the *Bam*HI site of a linearized vector remains possible. The cloned cDNA of interest can be easily excised by using the newly generated *Sma*I or *Hpa*II sites. This method allows *Bam*HI cloning even if the passenger cDNA contains one or more internal recognition sequences for this enzyme.

20.2 Linker Purity

Explain how you can verify the purity of a linker by using a labeling method involving polynucleotide kinase.

20.3 Linker Sites

Find the three restriction sites contained in the linker CATCGATC, and the five restriction sites in the linker GTGGCCAC.

20.4 Hexameric Linkers

Find the other tetrameric recognition sites in the *Sac*I linker GAGCTC. What is the advantage of hexameric linkers for the expression of cDNA molecules?

20.5 Linker Manipulations

Show how you can obtain an nine-nucleotide-long *Eco*RI linker from the octomeric d(GGAATTCC) and the decameric d(CCGAATTCGG) synthetic oligonucleotides.

20.6 Insertion of a *Hpa*II Fragment in a Vector Linearized With *Eco*RI

Show that, by using a polymerase and the *Eco*RI linker d(GGAATTCC), you can modify a cDNA fragment digested with *Hpa*II in order to insert it into a vector linearized with *Eco*RI.

20.7 Adaptor Sites

What are putative and encoded restriction sites in the adaptors 5′ − d(AATTCGCG) − 3′ and 5′ − d(GATCCCCGGG) − 3′?

20.8 Use of a Single-Stranded Adaptor

Show how an *Hha*I-digested cDNA can be ligated to a vector linearized with *Eco*RI by making use of the single-stranded adaptor 5′ − AATTCGCG − 3′.

20.9 Simultaneous Use of Two Adaptors

Show that by using the 5′ — AATTCCCGGG — 3′ adaptor on a DNA extremity treated with *Eco*RI, and the 5′ — TCGACCCGGG — 3′ on a DNA extremity digested with *Sal*I, you can generate a new restriction site.

20.10 Adaptor-Mediated cDNA Cloning

Explain how you can insert a *Bam*HI cDNA into a vector linearized by *Eco*RI by making use of the adaptors d(AATTCTCGAG) and d(GAGCTCCTAG). What new site do you generate and what is its advantage?

Part VI
Libraries

21

Genomic Libraries

Before in vitro encapsidation procedures became routine, it was typical to clone a cDNA of interest after enriching for its specific sequence. For instance, Tilghman et al. (1977) digested mouse genomic DNA with *Eco*RI, separated the fragments by preparative electrophoresis or reverse-phase chromatography, and probed the different fractions with a globin cDNA. One of the fractions (enriched 500 times) was cloned into λWES.λB. Out of 4,300 plaques tested by transfection, only three positive clones were obtained.

The present strategy consists of randomly cloning the genomic DNA of a given organism into bacteriophage vectors, thus creating a *genomic library.* The desired gene is isolated from the pool by different selection techniques. The methods used to generate the library guarantee that there is no specific discrimination against a particular DNA sequence. What at least one copy of every possible sequence is present in the library, it is known as *complete.* The exact probability for the presence of a given DNA sequence in the bank can be calculated by using:

$$N = \frac{\ln(1 - P)}{\ln(1 - f)}$$

where N is the number of recombinants and f the fraction of the genome in a given recombinant (Clarke and Carbon, 1976).

21.1. Example: Construction of a Human Genomic Library

Figure 21.1 illustrates the construction of the first human genomic library (Maniatis et al., 1978). The extracted eukaryotic DNA is first randomly fragmented. Ideally, the genomic DNA should be cut by mechanical techniques; however, the resulting fragments must be manipulated to display cohesive ends compatible with those of the linearized vector. Consequently, the fragments are generally obtained by partial digestion of the genomic DNA with restriction enzymes recognizing tetrameric sequences. (Hence they cut frequently.) The digested DNA is then frationated and molecules of approximately 20 kbp are pooled.

Cloning is performed into a λ derivative which can, as a replacement vector, integrate passenger DNA molecules up to 20 kbp long. Upon digestion of the phage with the appropriate restriction enzymes, the central domain is discarded and the fragmented genomic DNA is ligated to the purified phage arms. Recombinant bacteriophages are amplified following infection of *E. coli* and a library displaying high titers for several years is thus obtained.

21.2. Steps Involved in the Construction of a Genomic Library

- *Fractionation of the passenger DNA.* High-molecular-weight DNA is generally extracted under conditions that minimize its enzymatic degradation or cleavage through mechanical stress. The next step is the isolation of fragments that are neither too big nor too small to be inserted into phage replacement vectors (i.e., between 18 and 22 kbp for most vectors).

  Since they are compatible with *Bam*HI overhangs, the restriction enzymes *Sau*3A and *Mbo*I have been widely used in the construction of genomic libraries. These enzymes recognize the tetrameric sequence GATC and cut on the average every $4^4 = 256$ bp. In order to generate 20-kbp fragments, it is thus necessary to digest the genomic DNA incompletely, about once every 80 sites. The DNA fragments of the required size are isolated from the digestion mixture by centrifugation on a sucrose or sodium chloride gradient or by electrophoresis on agarose gels.

- *Removal of the central fragment of the replacement vector.* To ensure that most of the recombinant phages contain fragments of the genomic DNA, it is necessary to eliminate the central fragment of the replacement vector used in the cloning step. This "stuffer" is generally removed by elution, extraction from an agarose gel, or fractionation on a sucrose or sodium chloride gradient.

- *Ligation conditions.* Since the concatemeric form of recombinant phages is the best substrate for in vitro encapsidation, the ligation conditions are usually chosen to favor their formation. This goal is achieved by incubating the phage arms at 42 °C for 1 h in the presence of 10 mM $MgCl_2$ in order to anneal the *cos* sites of the cleaved vectors. The subsequent ligation step is performed

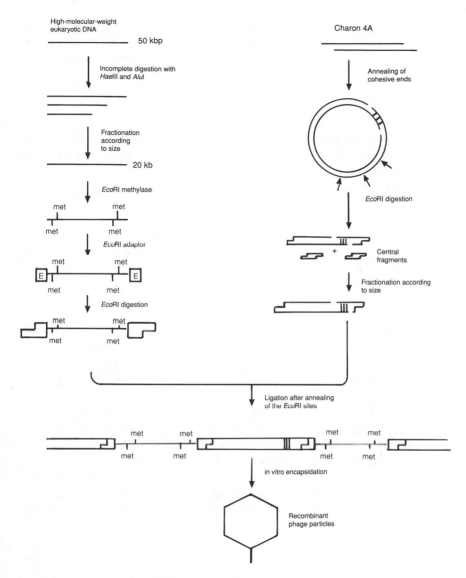

Figure 21.1. *Approach used by Maniatis for the construction of a genomic library.* The long-strand DNA is partially digested with *Hae*II and *Alu*I and 20-kbp fragments are selected by fractionation. The *Eco*RI methylase is used to methylate internal *Eco*RI sites, and synthetic *Eco*RI linkers are ligated to the blunted DNA molecules. The modified DNA is then digested with *Eco*RI. The replacement vector Charon 4A is circularized and digested with *Eco*RI, and the small central fragments (7 and 8 kbp) are removed by size fractionation. The altered genomic DNA and cloning vector are mixed and ligated to yield concatemeric molecules that are encapsidated in vitro using phage particles purified on CsCl gradient. About 10^5–10^6 clones are obtained per μg of DNA using this technique.

with relatively high DNA concentrations to favor intermolecular ligation reactions. In other words, DNA concentrations must be chosen so that $j \ll i$ for both the phage arms and the insert. The value of j can be calculated in extremities per ml using the formula:

$$j = \frac{6.1 \times 10^{22}}{mw^{3/2}} \tag{1}$$

Since the arms of the phage account for 31 kbp and the size of the insert is known, j values can be obtained. Parameter i represents the total DNA concentration in the reaction mixture (i.e., i of the phage arms + i of the insert), and may be calculated in ends per ml using:

$$i = 2 \cdot (6.02 \times 10^{23}) \cdot (M \cdot 10^{-3}) \tag{2}$$

Formation of concatemers is favored when $i = 10\,j$. Taking into account the fact that there are two phage arms per insert, the molar concentration of the insert and the phage arms can be determined using the above formulae.

21.3. EMBL Vectors

The EMBL vectors (Frischauf et al., 1983) are replacement phages that have the advantage of allowing the insertion of DNA fragments close to the theoretical limit (23 kbp). They are easy to use for the construction of genomic libraries.

The first vector of the series, EMBL1, was derived from λ1059 by substituting a central 5.7-kbp fragment corresponding to pBR322 DNA by the *trp*E gene from *E. coli*. In EMBL 2, the *Eco*RI sites of EMBL1 were removed. EMBL3 and 4 (Fig. 21.2) were constructed from EMBL2 by replacing the *Bam*HI sites by a multiple cloning site obtained from pUC7 (Fig. 21.3). In the case of EMBL3, sites *Sa*lI, *Bam*HI, and *Eco*RI are available on the 5′ side, and sites *Eco*RI, *Bam*HI, and *Sa*lI on the 3′ side. (EMBL4 contains the same sites in the opposite location.)

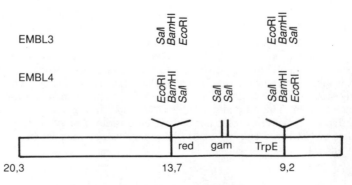

Figure 21.2.   *Simplified map of the EMBL3 and EMBL4 vectors.*

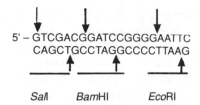

5' – GTCGACGGATCCGGGGAATTC
CAGCTGCCTAGGCCCCTTAAG

Sal   BamHI   EcoRI

Figure 21.3. *Multiple cloning site located on the 5' side of the central fragment of EMBL3.*

The most widely used site in these vectors is *Bam*HI (Fig. 21.4) since its sequence 5' — GATC — 3' is compatible with the overhangs generated by partial digestion of the genomic DNA with *Mbo*I or *Sau*3A.

The religation of the central fragment of EMBL3 can be prevented by enzymatic reactions. The full-length phage is first cleaved with *Bam*HI and then with *Eco*RI. As a result, three types of molecules are obtained: the arms of the phage exposing *Bam*HI overhangs, the central fragment flanked by *Eco*RI sites, and short *Bam*HI-*Eco*RI oligonucleotides. The latter can be readily separated from the larger DNA molecules by treatment with isopropanol since they are too small to be precipitated. Since the central fragment is no longer

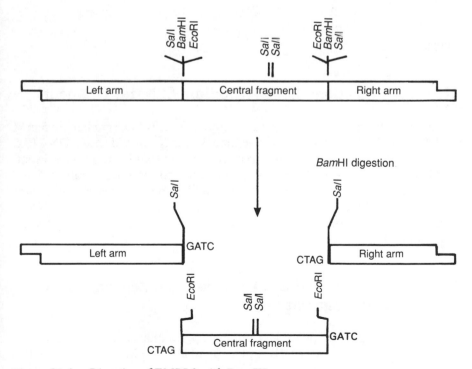

Figure 21.4. *Digestion of EMBL3 with BamHI.*

compatible with the phage arms, it will not compete with the cleaved genomic DNA during the ligation step, and only recombinant phages will be obtained. If this technique is not used, recombinant can still be selected by subjecting the religated mixture to *Sal*I digestion in order to cleave the reconstituted wild-type EMBL3 and prevent its encapsidation. It is of course necessary that no internal*Sal*I sites be present in the passenger DNA. If this is the case, the passenger DNA must first be modified by methylating its *Sal*I sites.

Recombinant phages generated with EMBL vectors are *gam⁻* and display an Spi- phenotype. This property allows their genetic selection on P2 lysogens. Passenger DNA molecules obtained following *Mbo*I digestion of the genomic DNA can be treated with alkaline phosphatase to prevent them from annealing to each other and generating hybrid inserts. As previously mentioned, this problem is generally solved by selecting those inserts ranging between 15 and 20 kbp on sucrose gradients.

Phage λ2001 is another derivative of λ1059 (Karn et al., 1980) that is similar to EMBL3 except that it contains unique cloning sites for the enzymes *Xba*I, *Sac*I, *Xho*I, *Bam*HI, *Hin*dIII, and *Eco*RI. These sites can accommodate DNA inserts ranging between 9 and 23 kbp and *spi* selection can also be used.

Another series of high-capacity cloning vectors has been described by Sorge et al. (1987). These vectors (e.g., λDASH) have all the advantages of the EMBL vectors but also encode the bacteriophage T3 and T7 promoters adjacent to the multiple cloning sites.

EXPERIMENTS

21.1. DNA Fractionation on a Sodium Chloride Gradient

Using a gradient marker at 1 ml per min, pour a 5–25% NaCl gradient in 5 mM EDTA, pH 8.0, in a 12.5-ml centrifuge tube. Layer 200 μl of digested DNA (0.5 μg/μl final concentration) at the top of the gradient and spin in a Beckman SW41 rotor or equivalent for 5 h at 35,000 rpm and room temperature. Fractionate into 250-μl aliquots and analyze a small amount on agarose gels. Store the remainder of the samples at − 20 °C. The fractions containing the DNA of required size can be ethanol precipitated, dried, and resuspended in the desired amount of TE buffer.

21.2. Preparation of the Phage Arms and Removal of the Central Fragment

Dilute 250 μg of purified phage DNA at a 0.5 μg/μl final concentration in ligation buffer (125 μl of 1 M Tris-HCl, pH 7.5, 50 μl of 0.5 M MgCl₂, 50 μl of 0.5

M β-mercaptoethanol, 125 μl of 20 mM ATP, and 150 μl ddH$_2$O). Add T4 DNA ligase and incubate at 16 °C for 2 h. The viscosity of the solution will increase as circularization of the phage progresses. Digest the ligated phage with *Eco*RI and remove the contaminating proteins by phenol and phenol–isoamyl alcohol (24:1) extraction. Precipitate the DNA with ethanol, dry the pellet, and resuspend in 1 ml of TE. Heat the solution at 68°C for 10 min and layer on top of a sucrose gradient. Phenol-extract the gradient fractions as above and resuspend in 250 μl TE after isopropanol precipitation of the DNA. Run 5 μl of each fraction on an 0.8% agarose gel and identify those containing the phage arms. These fragments can be extracted at need by preparative electroelution.

21.3. Ligation of Passenger DNA to the Phage Arms

Set up pilot reactions with 2:1, 1:1, and 0.5:1 ratios of insert to phage DNA to test the efficiency of the reaction. For each reaction use 2.5 μg of DNA resuspended in 10 μl of ligation buffer and incubate in the presence of T4 DNA ligase overnight at 14 °C. A control ligation with no insert is useful to determine the background level of undigested bacteriophage in the purified arms.

PROBLEMS

21.1 Ideal Genomic Library

What are the characteristics of an ideal genomic library?

21.2 Number of *Eco*RI Recombinants to Obtain a Given Sequence

The human genome consists of 2.8 $\times$ 10^6 kbp. How many recombinant phages must be generated to obtain at least one copy of any given sequence following *Eco*RI digestion?

21.3 Number of 20-kbp Fragments to Obtain a Complete Library

How many recombinant phages carrying 20-kbp inserts must be generated to obtain a complete human DNA genomic library?

21.4 Probability of Obtaining a Complete Genomic Library

How many recombinant phages are necessary to reach a 95% probability of presence of 20-kbp sequences in a human genomic library? Same question for a probability of 99%.

21.5 Complete Genomic Libraries in Different Organisms

How many clones are needed to obtain a complete (99% probability of presence of a 20-kbp sequence) genomic library in *E. coli* (genome size 4.2×10^6 bp), *Saccharomyces cerevisiae* (genome size 1.4×10^7 bp), and *Drosophila* (genome size 1.4×10^8 bp).

21.6 Lower Theoretical Limit for the Construction of a Genomic Library

What is the minimal amount of DNA necessary to construct a genomic library?

21.7 Advantages of EMBL Vectors

What are the three main advantages of the EMBL series of cloning vectors?

cDNA Libraries

Figure 22.1 shows a schematic representation of the steps involved in the construction of a *cDNA library* from polyadenylated RNA (adapted from Efstratiadis and Villa-Komaroff). The main steps consist of:

- Copying the poly(A) RNA into single-stranded cDNA
- Synthesizing double-stranded from single-stranded cDNA
- Linearizing pBR322 with *Pst*I
- Adding poly(dC) tracts to the 3′ ends of the cDNA and poly(dG) tracts to the 5′ ends of the digested vector
- Ligating the passenger cDNA into the modified vector
- Transforming an appropriate strain of competent *E. coli* with the resulting recombinant plasmids
- Selecting the bacterial clones for tetracyclin resistance and ampicillin sensitivity

The large amount of Tc^rAp^s recombinants obtained by this method (about 10,000) have different cDNA inserts and are known as a *cDNA library*. They are stored frozen at $-20\,°C$ on microplates in the presence of 50% glycerol.

22.1. Requirements for a Complete cDNA Library

Typical mammalian cells contain from 10 to 30,000 different messenger molecules. Depending on the tissue and the cell type, certain mRNA molecules

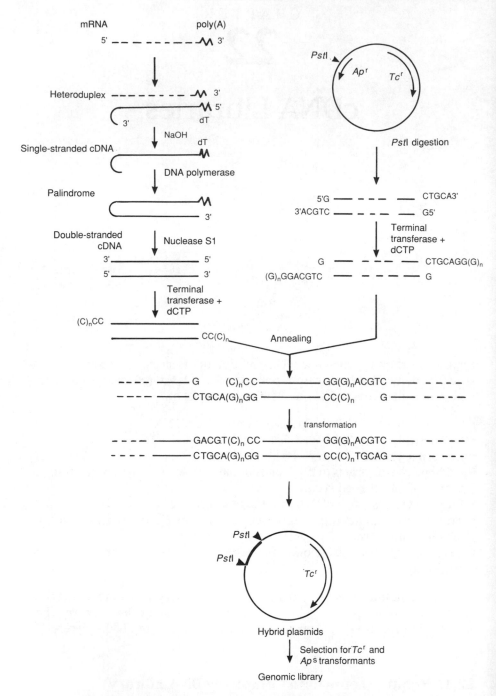

Figure 22.1. *Main steps involved in the construction of a cDNA library.*

Table 22.1   Abundance Of mRNA vs. Number Of Different mRNA Molecules In Mouse Liver (Adapted From Young et al., 1976)

No. Different mRNA Molecules	No. Copies Per Cell
9	12,000
700	300
11,500	15

may be present in large amounts while only a few copies of others are made (Table 22.1).

As a consequence, a cDNA library will only be complete if it also includes the low-abundance mRNA molecules (i.e., mRNAs present at less than 15 copies per cell). Theoretical calculations suggest that about 10^5 recombinants are sufficient to achieve a good representation of the rare mRNA molecules. This apparently high number of clones can be readily obtained since treatment of 1 μg of cDNA by homopolymeric tailing or addition of junction molecules generally yields from 1 to 5 $\times$ 10^5 clones.

22.2. Eliminating the S1 Nuclease Step

As previously mentioned, the use of S1 nuclease to remove the hairpin loop of double-stranded cDNA presents a number of inconveniences. In addition to partially digesting the 5′ ends of cDNA molecules, this enzyme increases cDNA loss, which results in a lower number of cDNA clones per μg of mRNA.

An alternative method for cDNA cloning was described by Okayama and Berg (1982) and is illustrated in Fig. 22.2. Briefly, the cloning vector is prepared to display a poly(T) extension that can anneal to the poly(A) tails of the mRNA. A single-stranded cDNA is then generated and its extremity is elongated with dC. An adaptor molecule containing a poly(G) tail and a restriction site compatible with one found on the plasmid (HindIII in the case of Fig. 22.2) are then used to circularize the recombinant plasmid, which now contains a hybrid cDNA/mRNA insert. Removal of the mRNA strand and synthesis of the complementary cDNA is achieved by using a combination of Rnase H (Appendix 23), DNA polymerase I and DNA ligase. The resulting plasmid is then introduced into competent *E. coli* cells.

A simplified method to obtain full-length cDNA clones was designed by Gubler and Hoffman (1983) and is depicted in Fig. 22.3. The first cDNA strand is synthesized by the classic approach of oligo(dT)/reverse transcriptase treatment. The synthesis of the complementary cDNA strand is inspired from the Okayama and Berg technique. The hybrid mRNA/cDNA molecule is treated with Rnase H, which generates nicks and gaps in the mRNA strand. The result-

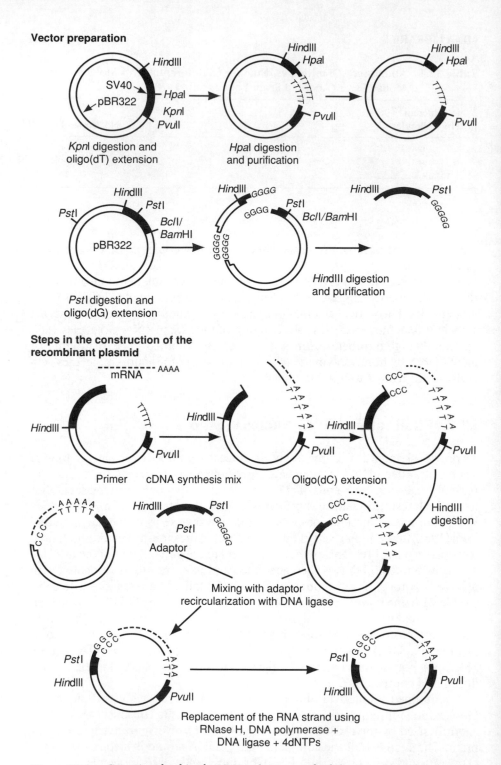

Figure 22.2. *Steps involved in the cDNA cloning method developed by Okayama and Berg.* The preparation of the vector (elongation poly[T]) and adaptor (elongation poly[G]) is also shown. The pBR-derived DNA is shown in white and the SV40-derived DNA in black.

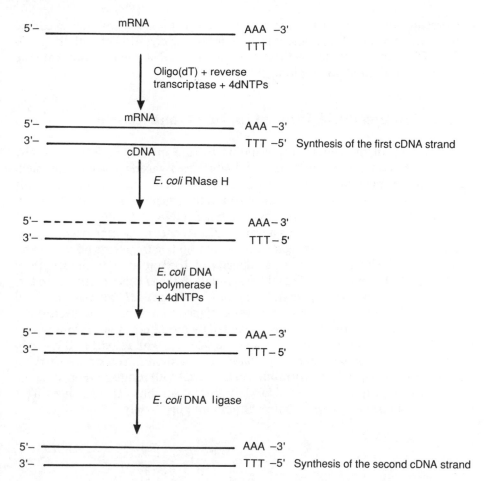

Figure 22.3. *Steps involved in the synthesis of full-length cDNA.*

ing primers are used by DNA polymerase I to generate the second cDNA strand. The process is completed by the ligation with T4 DNA ligase, or *E. coli* DNA ligase. (The latter enzyme is incapable of ligating double-stranded DNA molecules to each other and thus reduces the number of artifacts.) This method yields full-length cDNA molecules and results in high cloning efficiencies (up to 10^6 recombinants per μg of mRNA). It is, at present, the most widely used technique to generate cDNA clones. Okayama and Berg (1983) have modified pSV into pcDV1 by inserting an SV40-derived polyadenylation fragment in the vicinity of the cloning site. A linker fragment encoding the promoter of the early region of SV40 and two introns (to allow the cut of the transcribed RNA) were used to circularize the construct, which can express globin α and dihydrofolate reductase in mammalian cells.

Helfman et al. (1983) have shown that pUC8 could be efficiently used as an expression vector for cDNA clones. For instance, the cDNA of interest can be fitted with *Sal*I and *Eco*RI adaptors and, following digestion of both the modified cDNA and the vector by these enzymes, the cDNA can be ligated into the vector in the desired orientation.

22.3. Cloning cDNA Into λgt10, λgt11, and λZAPII

The cloning of the genes corresponding to rare mRNA molecules usually requires the construction of large cDNA libraries. By using phage vectors such as λgt10 and λgt11, libraries consisting of 10^5–10^7 clones may be obtained. This high number of recombinants is the result of the efficiency and reproducibility of the in vitro encapsidation process of λ phage DNA in *E. coli*.

λgt10 (*imm*^{434}*b*527) contains a unique *Eco*RI site in the gene encoding the λ repressor (Fig. 22.4). This vector can accommodate fragments up to 7.6 kbp in length, which corresponds to a capacity of 105% relative to the length of wild-type λ. The insertion of cDNA fragments in the *cI* gene results in the formation of *cI*$^-$ phages that give clear plaques (while λgt10 *cI*$^+$ give rise to turbid plaques). The λgt10 cloning vector was designed to optimize the insertion of DNA fragments shorter than those accepted by most known *imm* cloning vectors. Since most encapsidation extracts are selective with respect to the size of the phage, λgt10 grows vigorously. Plating of λgt10 on *E. coli* strains carrying the *high frequency of lysogeny* mutation (e.g., *hfl*A150) results in repressing the formation of plaques by parental (and thus nonrecombinant) *cI*$^+$ phages. This property can be used to select for recombinant phages only.

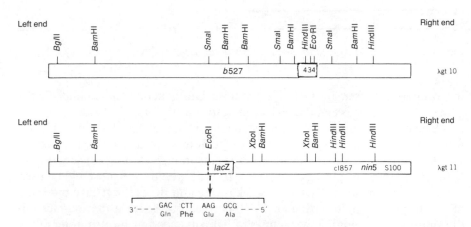

Figure 22.4. *Maps of λgt10 and λgt11.* Top: the *b*527 deletion in λgt10 removes the DNA sequences located between 49.1 and 57.4% of wild-type λ genome. The substitution *imm*^{434} replaces the wild-type λ DNA sequences located between 72.9 and 79.3%. Bottom: the position of the *lacZ* fragment in λgt11 is shown.

   Bacteriophage λgt11 (*lac5 cI857 nin5 S*100) is an expression vector constructed by Young and Davis (1983a,b). Cloning is performed at the unique
*Eco*RI site located 53 bp upstream of the translational termination codon of
the *lacZ* gene (Fig. 22.4). Approximately 7.2 kbp of DNA can be inserted at
this position. λgt11 encodes a temperature-sensitive repressor (*cI*857) that is
inactivated at 42 °C and an *amber* mutation (*S*100) that prevents the phage
from lysing host cells that do not carry a suppressor for this mutation (e.g.,
*sup*F). Foreign cDNA sequences cloned into λgt11 may be expressed as fusion
proteins to β-galactosidase (the *lacZ* gene product). Figure 22.5 illustrates the

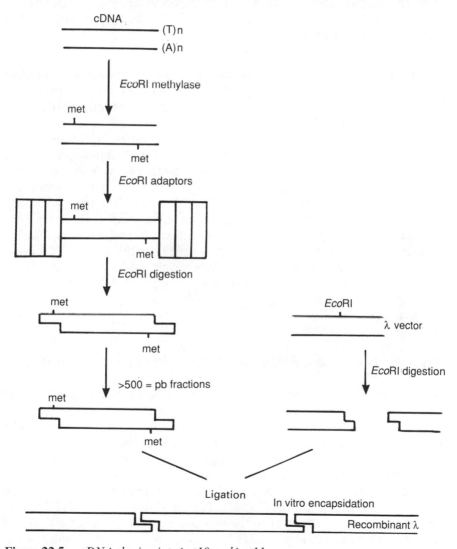

Figure 22.5.   *cDNA cloning into λgt10 and λgt11.*

steps used for cloning into λgt10 or λgt11. Detailed maps of the two vectors are provided in Appendix 21. The cloning process involves the following steps:

- Methylation of the double-stranded cDNA with the EcoRI methylase in order to protect internal sites from undesired cleavage
- Ligation of synthetic adaptors to the extremities of the cDNA following a brief treatment with DNA polymerase I, and digestion with EcoRI to eliminate the excess adaptors
- Chromatography on a BioGel A50m column to separate the modified DNA from the adaptors and fractionate it with respect to size (e.g., >500 bp)
- Digestion of the bacteriophage vector with EcoRI and ligation to the cDNA fragments of appropriate size
- In vitro encapsidation of the ligation products

If λgt10 is used as a cloning vector, recombinant phages are selected from clear plaques. In the case of λgt11, colored plaques are selected when the phages are plated on a $sup^F lacI^Q$ host in the presence of the lac operon inducer IPTG and the indicator substrate X-gal.

Another useful cloning vector for the construction of cDNA libraries is λZAPII (Short et al., 1988). It can accommodate up to 10 kbp of foreign DNA at six unique cloning sites located in the N-terminal region of the lacZ gene, which facilitates the blue/white selection of recombinants. It also encodes the T3 and T7 promoters for easy transcription of the insert. The main advantage of λZAPII is that the cloned cDNA molecules can be directly excised from the vector and recircularized into a λZAPII-encoded plasmid by using a *helper phage*. In this fashion, the additional steps consisting of the isolation of the passenger DNA and subcloning into an appropriate plasmid are avoided.

EXPERIMENTS

22.1. Synthesis of a Full-Length Complementary Strand

Synthesize the first cDNA strand in a reaction mixture containing 0.1 μg/μl of poly(A) RNA for a final volume of 100 μl. Transfer 30 μl to a fresh Eppendort tube and add 25 μl of 4× polymerase I buffer (400 μl HEPES, pH 7.6; 16 μl 1 M MgCl$_2$; 4.4 μl β-mercaptoethanol; 270 μl 1 M KCl; and 310 μl ddH$_2$O), 2.5 μl of 5× dNTP stock solution (containing each deoxynucleotide at a final concentration of 5 mM; each nucleotide is first prepared separately at a 20 mM concentration in 10 mM Tris-HCl, pH 7.5, and kept frozen at −80 °C), 1 μl of 15 mM β-NAD, and 10 μCi of α[^{32}P]dGTP. Add 30 units of E. coli DNA polymerase and 1 unit of RNase H to the reaction mixture per μg of cDNA and adjust the volume to 100 μl with ddH$_2$O. Incubate at 14 °C for 1 h and at room temperature for another hour.

Measure the amount of incorporated radioactivity using a 2-μl aliquot treated with 5 μl of 500 mM EDTA (pH 8.0), 2 μl 10% SDS, 50 μl of chloroform, and 50 μl of phenol previously equilibrated with 100 mM Tris-HCl, pH 8.0. Phenol extract the DNA by thoroughly mixing and centrifuging, recover the top aqueous phase, and reextract the phenol phase with 50 μl of TE. Pool the aqueous phases and count an aliquot. For direct visualization, use an aliquot corresponding to about 5,000 cpm; add 2 μg of carrier DNA and ethanol precipitate. Resuspend the dried pellet in 20 μl of alkaline buffer (30 mM NaOH, 1 mM EDTA) and add running buffer. Electrophorese on alkaline 1.2% agarose gel for 3 h using a recirculating buffer. Transfer the DNA onto a Whatman DE81 filter, dry, and expose.

The double-stranded cDNA is recovered following phenol extraction by addition of 150 μl of 4 M ammonium acetate and 600 μl ethanol followed by ethanol precipitation at -70 °C.

22.2. Using λgt10 and λgt11

- *Removal of the excess adaptors and fractionation on BioGel A50m.* Pour a 300 × 20 mm column using BioGel A50m in 10 mM Tris-HCl (pH 7.5), 100 mM NaCl, 1 mM Na$_3$EDTA. Calibrate the column using 5'-labeled *Hinf*I pBR322 fragments suspended in 10 μl of TE (pH 7.5), 1 ml of 0.25% bromophenol blue, and 50% glycerol. Adjust the flow rate so that the dye front migrates 1 cm in 10 min. Collect fractions, separate on a 1% agarose gel, and autoradiograph to obtain a calibration curve.

    Layer the cDNA on the column using three drops of about 15 μl each under the same conditions. Collect 1.5-ml fractions, count the ^{3}H radioactivity, and obtain an elution profile.
- *EcoRI digestion of λgt10 or λgt11.* Dilute 5 μg of vector DNA in 40 μl of ddH$_2$O; add 5 μl of 10× *Eco*RI buffer (500 mM Tris-HCl, pH 7.5; 1 M NaCl; 100 mM MgCl$_2$) and 15 units of *Eco*RI; adjust the volume to 50 μl with ddH$_2$O. Incubate the reaction mixture for 30 min at 37 °C, add another 15 units of enzyme, and incubate for an additional 30 min at the same temperature. Stop the digestion by addition of 1 μl of Na$_4$EDTA and heating at 70 °C for 10 min.
- *Ligation of the cDNA and the digested vector.* Pool the *Eco*RI-digested cDNA fractions of the desired size and centrifuge for 15 min to eliminate insoluble contaminants. Transfer the supernatant to a fresh tube and add 1 μg of *Eco*RI-digested λ DNA. Ethanol precipitate the cDNA, rinse the pellet with 70% ethanol, and resuspend in 4 μl of 10 mM Tris-HCl, pH 7.5, 10 mM MgCl$_2$. Incubate at 42 °C for 15 min. Add 0.5 μl of 10 mM ATP, 0.5 μl of 100 mM DTT, 0.1 μl of 1.6 mg/ml T4 DNA ligase and ligate for 2–16 h at 12–14 °C.

PROBLEMS

22.1 Importance of cDNA Libraries

What is the usefulness of cDNA libraries compared to genomic libraries? Can you suggest in which situations their construction is required?

22.2 Obtaining Full-Length cDNA Molecules

Is it easy to obtain full-length cDNA molecules? If not, can you suggest simple modifications of the classic technique that will lead to their obtention?

22.3 pUC Cloning Vectors

Explain why the pUC vectors are more widely used than the pBR series.

22.4 Length of the Complementary cDNA Strand

What is the length of the second strand of a cDNA molecule under conditions where full-length synthesis is complete? What is the difference before and after treatment with S1 nuclease?

22.5 λgt10 vs. λgt11

What are the criteria that determine whether λgt10 or λgt11 should be used?

23

Cosmid Libraries

Cosmids (Collins and Brüning, 1978; Collins and Hohn, 1979) are small plasmids (generally in the 5-kbp range) in which the bacteriophage λ *cos* site and adjoining sequences have been cloned. Since the only elements necessary for bacteriophage λ encapsidation are the *cos* sites, its neighboring sequences, and the presence of a DNA insert of 37–52 kbp between two consecutive *cos* sites, cosmids carrying inserts can also be encapsidated. Thus, cosmid vectors can accommodate passenger DNA in the 32–47-kbp range, which is a much larger fragment than what can be cloned in bacteriophage λ. The term cosmid was coined to emphasize the hybrid nature of these cloning vectors that display the characteristics of both plasmids and bacteriophages since they can be encapsidated in vitro.

   Figure 23.1 shows, as an example, the map of cosmid pJB8, which displays the following features:

- pJB8 size 5.4 kbp.
- The cosmid encodes the *cos* site and its flanking sequences.
- The plasmid-derived origin of replication is *ori*.
- pJB8 encodes the *amp* gene encoding β-lactamase which confers ampicillin resistance to the transformed bacteria.
- Foreign DNA fragments can be cloned in the unique *Bam*HI site and can be excised by making use of the flanking *Eco*RI sites.

The first generation of cosmids carried a small number of cloning sites and only one drug resistance marker; because of their small size, the cosmid features could not be used to their full extent. Therefore, an improved series of cosmids

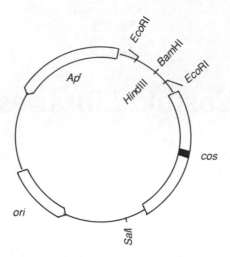

Figure 23.1. *Map of cosmid pJB8* (from Ish-Horowicz and Burke, 1981).

was constructed. As an example, cosmid MuA-3 encoded a *cos* site in a 403-bp *Hinc*II λ fragment cloned in the *Pst*I site of pBR322. The MuA-10 derivatives were further improved by introducing a second *Pst*I site. Cosmid Homer I (Chia et al., 1982) contains a 1.78-kbp *Bgl*II fragment encoding a *cos* site derived from the circularized form of Charon 4A into the *Bam*HI site of pBR322. As a result, the tetracycline resistance gene is inactivated. In addition, cosmids Homer V and VI encode a *Tn*5 insert for the bacterial aminoglycide phosphotransferase (APH 3′) gene. The latter can be selected on a derivative of gentamycin (G418, Wolfe et al., 1984). In Homer V, the APH 3′ gene is under transcriptional control of the thymidine kinase promoter from the herpes simplex virus. Furthermore, both Homer V and VI carry a *Hpa*II–*Hin*dIII fragment encoding the SV40 origin of replication. In another cosmid, pGcos4, the marker is dihydrofolate reductase (DHFR), which gives resistance to methotrexate (Gitschier et al., 1984). Appendix 24 gives the characteristics of some important cosmids.

Cosmids are generally used for the cloning of large DNA fragments and are particularly helpful for the construction of genomic libraries in which passenger DNA molecules are obtained following partial digestion with specific restriction endonucleases.

Large DNA fragments can be readily obtained from genomic DNA through partial digestion with enzymes such as *Sau*3A (↓GATC). The recognition sequence of the latter endonuclease contains a base of each type and is therefore a random cutter unsensitive to the nucleotide composition of the genome. Fractionation of the partially digested DNA and recovery of the fragments ranging between 30 and 45 kbp is then accomplished by ultracentrifugation on sucrose gradient or preparative DNA electrophoresis on agarose gels.

Figure 23.2 summarizes the different steps used for the cloning of passenger DNA in a cosmid. Briefly, the vector linearized with *Bam*HI and the partially digested DNA of appropriate size are mixed and ligated. By using a high DNA concentration and an excess of vector DNA, the formation of concatemers through intermolecular ligations is favored. The ligation products are then

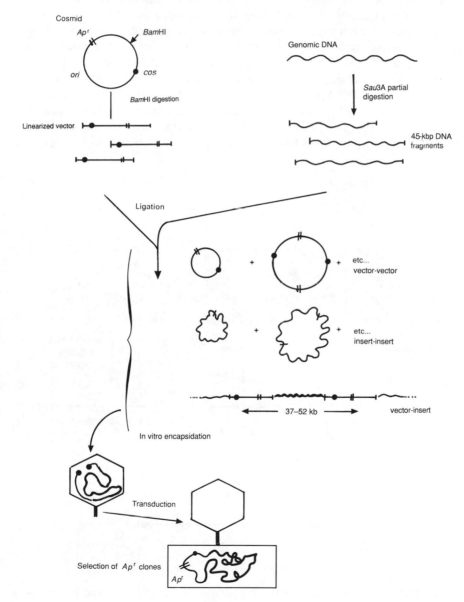

Figure 23.2.   *DNA cloning in a cosmid.*

encapsidated in vitro. The molecules containing two consecutive *cos* sites in the same orientation and separated by about 45 kbp of passenger DNA have the highest chance of being encapsidated. The resulting phage particles are used to infect an *E. coli* strain of the appropriate genotype, and the recombinant cosmids are selected on the basis of their resistance to an antibiotic (e.g., ampicillin in the case of pJB8).

The selection of large pieces of DNA prior to the ligation step prevents the formation of artefacts that may arise, for instance, when two or more small pieces of DNA, which were not normally adjacent in the genomic DNA, anneal and become ligated as one piece in the cosmid. In addition, as a result of large size selection, the genomic DNA is less likely to religate on itself. Finally, concatemers may be obtained through the formation of hybrids between passenger DNA and cosmid DNA, but can also result from the annealing of several cosmids into polycosmids. In order to reduce the tendency of the digested genomic DNA to religate to itself, cloning can be carried out using elevated concentrations of cosmid DNA. This has, however, the negative side effect of favoring

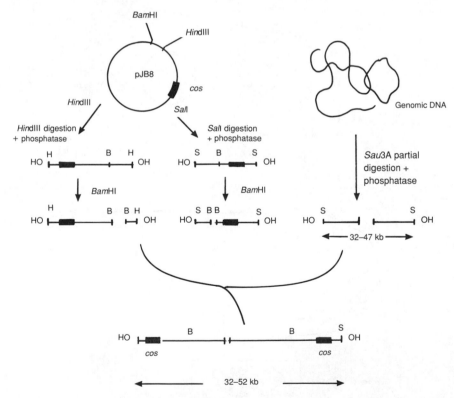

Figure 23.3. *Cloning of genomic DNA in pJB8* (from Ish-Horowicz and Burke, 1981).

the generation of polycosmids. In general, religation of the cosmid to itself is avoided by treating the linearized vector with alkaline phosphatase.

An alternative technique to prevent the ligation of the vector and passenger DNA to themselves as well as allowing the cloning of small, nonadjacent DNA fragments, was described by Ish-Horowicz and Burke (1981) and is illustrated in Fig. 23.3.

The cosmid (e.g., pJB8) is linearized with two different endonucleases and treated with alkaline phosphatase. Subsequent digestion with *Bam*HI generates a right and left phage arm, which are semiphosphorylated. The arms are next ligated with genomic DNA that has been partially digested with *Sau*3A and treated with alkaline phosphatase. Following ligation, only those hybrid molecules consisting of concatemers between right and left arms separated by about 50 kbp will become encapsidated and yield recombinants.

The main advantage of cosmid vector is that they make possible the cloning of large DNA fragments. This feature is essential for the construction of certain libraries. For instance, complete libraries of *Saccharomyces, Drosophila,* and mouse DNA require $4 \times 10^2, 4 \times 10^3$, and 7×10^4 recombinants, respectively, each containing 40-kbp inserts. The cloning efficiency in cosmids is about 10^3–10^6 transduced particles per μg of DNA, which is several orders of magnitude higher than that obtained with 40-kbp plasmids.

The choice of the bacterial strain used in the transduction process affects the stability of the cosmids and in general *rec*A hosts are used. A possible inconvenience to the use of cosmids for the creation of genomic libraries is related to the fact that smaller cosmids have a replicative advantage relative to larger ones. Over time, this may result in the loss of cosmids that have incorporated the largest genomic DNA fragments.

EXPERIMENTS

23.1. Digestion of Genomic DNA With *Sau*3A

The optimal digestion conditions are usually determined by making a serial dilution of the enzyme as follows: Aliquot 30 μl of DNA at 1 μg/μl in the appropriate restriction buffer in a fresh Eppendorf tube. Aliquot 15 μl of DNA in four additional tubes (numbered 2–5). Add 1 unit of *Sau*3A to the first tube, transfer 15 μl of the reaction mixture to tube number 2, and repeat with 3 to 5. Place the tubes at 37 °C and digest for 20 min. Stop all the reactions simultaneously by heating at 70 °C for 10 min. Resolve a 5-μl aliquot from each tube on a 0.5% agarose gel developed with a Tris–acetate buffer for 15 h at 25 V in order to determine optimal conditions. Use undigested λ DNA (50 kDa) as a molecular weight marker.

23.2. Isolation of Restriction Fragments About 45 kDa in Size

Separate the products of the partial digestion of the genomic DNA by *Sau*3A on a 0.5% low-melting-point agarose gel using a Tris-acetate buffer and 30 mA constant current at 40 °C for 20 h. Again, use undigested λ DNA as a marker. Identify the position of the DNA of interest following staining with ethidium bromide, excise the agarose slice corresponding to the appropriate molecular weight, and transfer to a fresh Eppendorf tube. Elute the DNA by adding five volumes of TE buffer and incubating at 68 °C for a few minutes. Recover the DNA by phenol extraction followed by ethanol precipitation and resuspend in the desired volume of TE.

23.3. Cosmid Preparation and Ligation

Digest the cosmid (e.g., pJB8) by mixing 25 µl of vector DNA (at 1 µg/µl) with 20 µl of 10× *Bam*HI buffer and 50 units of restriction enzyme. Adjust the volume to 200 µl using ddH$_2$O and transfer to a 37 °C incubator for 2 h. After checking that digestion is complete, extract the DNA once with phenol–chloroform and twice with chloroform. Ethanol precipitate the aqueous phase and resuspend the washed and dried pellet in 100 µl of 10 mM Tris-HCl, pH 8.0. Add 1 unit of alkaline phosphatase and incubate for 45 min at 37 °C. Reextract the DNA once with phenol–chloroform and twice with chloroform, ethanol precipitate, wash, dry, and resuspend in TE to achieve a final concentration of about 1 µg/µl.

For the ligation, mix 2 µl (2 µg) of genomic DNA partially digested with *Sau*3A with 2 µl (2 µg) of *Bam*HI-digested and alkaline-phosphatase-treated pJB8. Add 0.5 µl of 10× ligation buffer and 0.5 µl of T4 DNA ligase, and adjust the volume to 5 µl using ddH$_2$O. Incubate at 12 °C for 3 h.

PROBLEMS

23.1 Theoretical Cloning Capacity of Cosmids

What are the theoretical cloning capacities in cosmids 25, 11, and 3 kbp long?

23.2 Features of *Sau*3A

What are the features of the restriction enzyme *Sau*3A that make it particularly useful for the creation of cosmid genomic DNA libraries and cloning into cosmids?

23.3. Cosmid Preparation and Ligation

In the case of the human genome ($G = 2.8 \times 10^9$ bp), determine how many cosmid clones, N, of length $L = 40$ kbp are necessary to obtain a complete library at a probability $P = 0.99$? {Use the formula $P = 1 - [(1 - L/G)^N]$.} Same question for a phage library with $L = 15$ kbp.

APPENDIX
1

Commonly Used Bacterial Strains

Appendix 1   Commonly Used Bacterial Strains

Strain	Genotype	Comments
HB101	F$^-$ hsdS20 (r$_B^-$m$_B^-$) recA13 ara-14 proA2 lacY1 galK2 rspL20 (Sm^r) xyl-5 mtl-1 supE44 λ$^-$	Hybrid K-12 × B E. coli strain; used widely for plasmid amplification and transformation
RR1	Identical to HB101 but recA$^-$	Used for the high-efficiency transformation of cDNA elongated with homopolymeric tails
C600	F$^-$ thi-1 thr-1 leuB6 lacY1 tonA21 supE44 λ$^-$	Also known as CR34
LE392	F$^-$ hsdR514 (r$_K^-$m$_K^-$) supE44 supF58 lacY1 or Δ(lac YZI)6 galK2 galT22 metB1 trpR55 λ$^-$	su$^+$ strain usually used for propagating bacteriophage λ and its recombinants
X1776	F$^-$ tonA53 dapD8 minA1 glnV44 (supE42) Δ(gal,uvrB)40 minB2 rfb-2 gyrA25 thyA142 oms-2 metC65 oms-1 (tte-I) Δ(bioH-asd(29 cycB2 cyeA1 hsdR2 λ$^-$	Strain containing a large number of mutations as well as two deletions, which makes its growth impossible outside of a laboratory; it is recommended by the NIH for plasmid transformation and amplification
Q358	hsdR$_K^-$ hsdM$_K^+$ supF φ80^r	su$^+$ host used for λ1059 growth
Q359	hsdR$_K^-$ hsdM$_K^+$ supF φ80^r 92	su$^+$ host used for the detection of Spi$^-$ recombinants of bacteriophage λ
DH1	F$^-$ recA1 endA1 gyrA97 thi-1 hsdR17 (r$_K^-$m$_K^-$) supE44 λ$^-$	recA$^-$ host used for plasmid and cosmid transformation

Strain	Genotype	Comments
JM83	*ara lac pro thi strA* ϕ80 ΔlacZM15 ($r_K^+ m_K^-$)	Host for pUC plasmids
BHB2688	N205 *recA*$^-$ λ^r[λimm^{434} *c*I(*ts*) b2 *red*$^-$ *E*^{am} *S*^{am}]	λ lysogen used in the preparation of encapsidation extracts
BHB2690	N205 *recA*$^-$ λ^r[λimm^{434} *c*I(*ts*) b2 *red*$^-$ *D*^{am} *S*^{am}]	As above
C-1A		Wild-type *E. coli* C strain; the latter is F$^-$ and does not contain restriction or modification activities; this host is *su*$^-$ and used for complementation tests with bacteriophage λ containing *amber* mutations
CSH18	Δ(*lac, pro*) *sup*E *thi*$^-$ (F$'$, *lac*Z- *pro*A$^+$B$^+$)	Suppressor strain used for the selection of λ recombinants containing the *lac* gene

2

Some of the Known Restriction Enzymes

Appendix 2 Some Of The Known Restriction Enzymes

Microorganism	Enzyme	Recognition Sequence
Acetobacter aceti	*Aat*I	AGGCCT
	AatII	GACGT↓C
Acetobacter aceti sub. *liquefaciens*	*Aac*I	GGATCC
Acetobacter aceti sub. *liquefaciens*	*Aae*I	GGATCC
Acetobacter aceti sub. *orleanensis*	*Aor*I	CC↓(^A_T)GG
Acetobacter pasteurianus sub. *pasteurianus*	*Apa*I	GGGCC↓C
Achromobacter immobilis	*Aim*I	?
Acinetobacter calcoaceticus	*Acc*I	GT↓(^A_C)(^G_T)AC
	*Acc*II	CGCG
	*Acc*III	?
Actinomadura madurae	*Ama*I	TCGCGA
Agmenellum quadruplicatum	*Aqu*I	CPyCGPuG
Agrobacterium tumefaciens	*Atu*AI	? ·
Agrobacterium tumefaciens B6806	*Atu*BI	CC(^A_T)GG
Agrobacterium tumefaciens HBV7	*Atu*BVI	?
Agrobacterium tumefaciens ID 135	*Atu*II	CC(^A_T)GG
Agrobacterium tumefaciens C58	*Atu*CI	TGATCA
Alcaligenes (species)	*Asp*AI	G↓GTNACC
Anabaena catanula	*Aca*I	?
Anabaena cylindrica	*Acy*I	GPu↓CGPyC

Microorganism	Enzyme	Recognition Sequence
Anabaena flos-aquae	*Afl*I	G↓G(^A_T)CC
	*Afl*II	G↓TTAAG
	*Afl*III	A↓CPuPyGT
Anabaena oscillarioides	*Aos*I	TGC↓GCA
	*Aos*II	GPu↓CGPyC
Anabaena strain Waterbury	*Ast*WI	GPu↓CGPyc
Anabaena suscylindrica	*Asu*I	G↓GNCC
	*Asu*II	TT↓CGAA
	*Asu*III	GPu↓CGPyC
Anabaena variabilis	*Ava*I	C↓PyCGPuG
	*Ava*II	C↓G(^A_T)CC
	*Ava*III	ATGCAT
Anabaena variabilis^{uw}	*Avr*I	CPyCGPuG
	*Avr*II	CCTAGG
Aphanothece halophytica	*Aha*I	CC(^C_G)GG
	*Aha*II	?
	*Aha*III	TTT↓AAA
Arthrobacter luteus	*Alu*I	AG↓CT
Arthrobacter pyridinolis	*Apy*I	CC↓(^A_T)GG
Bacillus acidocaldarius	*Bac*I	CCGCGG
Bacillus amyloliquefaciens F	*Bam*FI	GGATCC
Bacillus amyloliquefaciens H	*Bam*HI	G↓GATCC
Bacillus amyloliquefaciens K	*Bam*KI	GGATCC
Bacillus amyloliquefaciens N	*Bam*NI	GGATCC
	BamN^x	G↓G(^A_T)CC
Bacillus aneurinolyticus	*Ban*I	GGPyPuCC
	BanII	GPuGCPy↓C
	BanIII	ATCGAT
Bacillus brevis S	*Bbv*SI	GC$\overset{*}{C}$(^A_T)GC
Bacillus brevis	*Bbv*I	GCAGC(8/12)
Bacillus caldolyticus	*Bcl*I	T↓GATCA
Bacillus centrosporus	*Bcn*I	CC(^C_G)↓GG
Bacillus cereus	*Bce*14579	?
Bacillus cereus	*Bce*1229	?
Bacillus cereus	*Bce*170	CTGCAG
Bacillus cereus Rf sm st	*Bce*R	CGCG
Bacillus globigii	*Bgl*I	GCCNNNN↓NGGC
	*Bgl*II	A↓GATCT
Bacillus megaterium 899	*Bme*899	?
Bacillus megaterium B205-3	*Bme*205	?
Bacillus megaterium	*Bme*I	?
Bacillus pumilus AHU 1387 A	*Bpu*I	?
Bacillus sphaericus	*Bsp*1286	?
Bacillus sphaericus R	*Bsp*RI	GG↓CC
Bacillus stearothermophilus C1	*Bst*CI	GGCC

Microorganism	Enzyme	Recognition Sequence
Bacillus stéarothermophilus C11	*Bss*CI	GGCC
Bacillus stearothermophilus G3	*Bst*GI	TGATCA
	Bst GII	CC(^A_T)GG
Bacillus stearothermophilus G6	*Bss*GI	CCANNNNNTGG
	*Bss*GII	GATC
Bacillus stearothermophilus H1	*Bst*HI	CTCGAG
Bacillus stearothermophilus H3	*Bss*HI	CTCGAG
	*Bss*HII	GCGCGC
Bacillus stearothermophilus H4	*Bsr*HI	GCGCGC
Bacillus stearothermophilus P1	*Bss*PI	?
Bacillus stearothermophilus P5	*Bsr*PI	?
	*Bsr*PII	GATC
Bacillus stearothermophilus P6	*Bse*PI	GCGCGC
Bacillus stearothermophilus P8	*Bso*PI	GATC
Bacillus stearothermophilus P9	*Bso*PI	?
Bacillus stearothermophilus T12	*Bst*TI	CCANNNNNTGG
Bacillus stearothermophilus X1	*Bst*XI	CCANNNNN↓NTGG
	Bst XII	GATC
Bacillus stearothermophilus 1503-4R	*Bst*I	G↓GATCC
Bacillus stearothermophilus 240	*Bst*AI	?
Bacillus stearothermophilus ET	*Bst*EI	?
	*Bst*EII	G↓GTNACC
	*Bst*EIII	GATC
Bacillus stearothermophilus	*Bst*PI	G↓GTNACC
Bacillus stearothermophilus	*Bst*NI	CC↓(^A_T)GG
Bacillus stearothermophilus 822	*Bse*l	GGCC
	*BS*II	GTTAAC
Bacillus subtilis strain R	*Bsu*RI	GG↓C̊C
Bacillus subtilis Marburg 168	*Bsu*M	?
Bacillus subtilis	*Bsu*6663	?
Bacillus subtilis	*Bsu*1076	GGCC
Bacillus subtilis	*Bsu*1114	GGCC
Bacillus subtilis	*Bsu*1247	CTGCAG
Bacillus subtilis	*Bsu*1145	?
Bacillus subtilis	*Bsu*1192	CCGG
	*Bsu*1192II	CGCG
Bacillus subtilis	*Bsu*1193	CGCG
Bacillus subtilis	*Bsu*1231I	CCGG
	*Bsu*1231II	CGCG
Bacillus subtilis	*Bsu*1259	?
Bifidobacterium bifidum	*Bbi*I	CTGCAG
	*Bbi*II	GPuCGpyC
	*Bbi*III	CTCGAG
	*Bbi*IV	?
Bifidobacterium breve	*Bde*I	GGCGC↓C
Bifidobacterium breve S1	?	?
Bifidobacterium breve S50	*Bbe*AI	GGCGCC
	*Bbe*AII	?

Microorganism	Enzyme	Recognition Sequence
Bifidobacterium infantis 659	*Bin*I	GGATC
Bifidobacterium infantis S76e	*Bin*SI	CC(^A_T)GG
	*Bin*SII	GGCGCC
Bifidobacterium longum E194b	*Blo*I	?
Bifidobacterium thermophilum RU326	*Bt*I	CTCGAG
	*Bth*II	?
Bordetella bronchiseptica	*Bbr*I	AGCTT
Bordetella pertussis	*Bpe*I	AAGCTT
Brevidacterium albidum	*Bal*I	TGG↓$\overset{*}{C}$CA
Brevibacterium luteum	*Blu*I	C↓TCGAG
	*Blu*II	GGCC
Calothrix scopulorum	*Csc*I	CCGC↓GG
Caryophanon latum L.	*Cla*I	AT↓CGAT
Caryophanon latum	*Clm*I	GGCC
	*Clm*II	GG(^A_T)CC
Caryophanon latum	*Clt*I	GG↓CC
Caryophanon latum RII	*Clu*I	?
Caryophanon latum H7	*Cal*I	?
Caulobacter crescentus CB	*Ccr*I	?
	*Ccr*II	CTCGAG
Caulobacter fusiformis	*Cfu*I	G$\overset{*}{A}$TC
Chloroflexus aurantiacus	*Cau*I	GG(^A_T)CC
	*Cau*II	CC↓(^C_G)GG
Chromatium vinosum	*Cvn*I	CC↓TNAGG
Chromobacterium violaceum	*Cvi*I	?
Citrobacter freundii	*Cfr*I	Py↓GGCCPu
Clostridium pasteurianum	*Cpa*I	GATC
Corynebacterium humiferum	*Chu*I	AAGCTT
	*Chu*II	GTPyPUOC
Corynebacterium petrophilum	*Cpe*I	TGATCA
Cystobacter velatus Plv9	*Cve*I	?
Desulfovibrio desulfuricans	*Dde*I	C↓TNAG
Sonche Norway	*Dde*II	CTCGAG
Desulfovibrio desulfuricans	*Dds*I	GGATCC
Diplococcus pneumoniae	*Dpn*I	G$\overset{*}{A}$↓TC
Diplococcus pneumoniae	*Dpn*II	GATC
Enterobacter aerogenes	*Eae*I	Py↓GGCCPu
Enterebacter cloacae	*Ecl*I	?
	*Ecl*II	CC(^A_T)GG
Enterobacter cloacae	*Eca*I	G↓GTNACC
	*Eca*II	CC(^A_T)GG
Enterobacter cloacae	*Ecc*I	CCGCGG
Escherichia coli pDXI	*Eco* DXI	ATCA(N)₇ATTC

Microorganism	Enzyme	Recognition Sequence
Escherichia coli J62 pLG74	*Eco*RV	GATAT↓C
Escherichia coli RY13	*Eco*RI	G↓AÅTTC
	*Eco*RI′	PuPuA↓TPyPy
Escherichia coli R245	*Eco*RII	↓CC̊(^A_T)GG
Escherichia coli B	*Eco* B	TGÅ(N)$_8$TGCT
Escherichia coli K	*Eco*K	AAC(N)$_6$GTGC
Escheria coli (PI)	*Eco*PI	AGÅCC
Escherichia coli P15	*Eco*P15	CAGCAG
Flavobacterium okeanokoites	*Fok*I	GGATG(9/13)
Fremyella diplisiphon	*Fdi*I	G↓G(^A_T)CC
	*Fdi*II	TGC↓GCA
Fusobacterium nucleatum A	*Fnu*AI	G↓ANTC
	*Fnu*AII	GATC
Fusobacterium nucleatum C	*Fnu*CI	↓GATC
Fusobacterium nucleatum D	*Fnu*DI	GG↓CC
	*Fnu*DII	CG↓CG
	*Fnu*DIII	GCG↓C
Fusobacterium nucleatum E	*Fnu*EI	↓GATC
Fusobacterium nucleatum 48	*Fnu*48I	?
Fusobacterium nucleatum 4H	*Fnu*4HI	GC↓NGC
Gluconobacter dioxyacetonicus	*Gdi*I	AGG↓CCT
	*Gdi*II	Py↓GGCCG
Gluconobacter dioxyacetonicus	*Gdo*I	GGATCC
Gluconobacter oxydans sub. melonogenes	*Gox*I	GGATCC
Haemophilus aegyptius	*Hae*I	(^A_T)GG↓CC(^A_T)
	*Hae*II	PuGCGC↓Py
	*Hae*III	GG↓C̊C
Haemophilus aphrophilus	*Hap*I	?
	*Hap*II	C↓CGG
Haemophilus gallinarum	*Hga*I	GACGC(5/10)
Haemophilus haemoglobinophilus	*Hhg*I	GGCC
Haemophilus haemolyticus	*Hha*I	GC̊G↓C
	*Hha*II	GANTC
Haemophilus influenzae GU	*Hin*GUI	GCGC
	*Hin*GUII	GGATG
Haemophilus influenzae 173	*Hin*173	AAGCTT
Haemophilus influenzae 1056	*Hin*10561	CGCG
	*Hin*1056II	?
Haemophilus influenzae serotype b, 1076	*Hin*bIII	AAGCTT
Haemophilus influenzae serotype c, 1160	*Hin*cII	GTPyPuAC
Haemophilus influenzae serotype c, 1161	*Hin*cII	GTPyPuAC
Haemophilus influenzae serotype e	*Hin*eI	CGAAT
Haemophilus influenzae R$_b$	*Hin*bIII	AAGCTT
Haemophilus influenzae R$_c$	*Hin*cII	GTPyPuAC

Microorganism	Enzyme	Recognition Sequence
Haemophilus influenzae R_d	*Hind*I	C$\overset{*}{A}$C
	*Hind*II	GTPy↓Pu$\overset{*}{A}$C
	*Hind*III	$\overset{*}{A}$↓AGCTT
	*Hind*IV	G$\overset{*}{A}$C
Haemophilus influenzae R_f	*Hinf*I	G↓ANTC
	*Hinf*II	AAGCTT
	*Hinf*III	CGAAT
Haemophilus influenzae H-1	*Hin*HI	PuGCGCPy
Haemophilus influenzae P_1	*Hin*P_1I	G↓CGC
Haemophilus influenzae S_1	*Hin*S1	GCGC
Haemophilus influenzae S_2	*Hin*S_2	GCGC
Haemophilus influenzae JC9	*Hin*JCI	GTPy↓PuAC
	*Hin*JCII	AAGCTT
Haemophilus parahaemolyticus	*Hph*I	GGTGA(8/7)
Haemophilus parainfluenzae	*Hpa*I	GTT↓$\overset{*}{A}$AC
	*Hpa*II	C↓$\overset{*}{C}$GG
Haemophilus suis	*Hsu*I	A↓AGCTT
Halococcus agglomeratus	*Hag*I	?
Herpetosiphon giganteus HP 1023	*Hgi*AI	G$\binom{A}{T}$GC$\binom{A}{T}$↓C
Herpetosiphon giganteus Hpg 5	*Hgi*BI	G↓G$\binom{A}{T}$CC
Herpetosiphon giganteus Hpg 9	*Hgi*CI	G↓GPyPuCC
	*Hgi*CII	G↓G$\binom{A}{T}$CC
	*Hgi*CIII	G↓TCGAC
Herpetosiphon giganteus Hpa 2	*Hgi*DI	GPu↓CGPyC
	*Hgi*DII	G↓TCGAC
Herpetosiphon giganteus Hpg 24	*Hgi*EI	G↓G$\binom{A}{T}$CC
	*Hgi*EII	ACC(N)$_6$GGT
Herpetosiphon giganteus Hpg 14	*Hgi*FI	?
Herpetosiphon giganteus Hpa 1	*Hgi*GI	GPu↓CGPyC
Herpetosiphon giganteus HP 1049	*Hgi*HI	G↓GPyPuCC
	*Hgi*HII	GPu↓CGPyC
	*Hgi*HIII	G↓G$\binom{A}{T}$CC
Herpetosiphon giganteus HFS 101	*Hgi*JI	?
	HgiJII	GPuGCPy↓C
Herpetosiphon giganteus Hpg 32	*Hgi*KI	?
Klebsiella pneumoniae OK8	*Kpn*I	GGTAC↓C
Mastigocladus laminosus	*Mla*I	TT↓CGAA
Microbacterium thermosphactum	*Mth*I	GATC
Micrococcus luteus	*Mlu*I	A↓CGCGT
Micrococcus radiodurans	*Mra*I	CCGCGG
Microcoleus species	*Mst*I	TGC↓GCA
	MstII	CC↓TNAGG
Moraxella bovis	*Mbo*I	↓GATC
	*Mbo*II	GAAGA

Microorganism	Enzyme	Recognition Sequence
Moraxella bovis	*Mbv*I	?
Moraxella glueidi LG1	*Mgl*I	?
Moraxella glueidi LG2	*Mgl*II	?
Moraxella kingae	*Mki*I	AAGCTT
Moraxella nonliquefaciens	*Mno*I	C↓CGG
	*Mno*II	?
	*Mno*III	GATC
Moraxella nonliquefaciens	*Mnl*I	CCTC(7-7)
Moraxella nonliquefaciens	*Mnn*I	GTPyPuAC
	*Mnn*II	GGCC
	*Mnn*III	?
	*Mnn*IV	GCGC
Moraxella nonliquefaciens	*Mni*I	GGCC
	*Mni*II	CCGG
Moraxella osloensis	*Mos*I	GATC
Moraxella phenylpyruvica	*Mph*I	CC($^{A}_{T}$)GG
Moraxella species	*Msp*I	C↓CGG
Myxococcus stipitatus Mxs2	*Msi*I	CTCGAG
	*Msi*II	?
Myxococcus virescens V-2	*Mvi*I	?
	*Mvi*II	?
Neisseria caviae	*Nca*I	GANTC
Neisseria cinerea	*Nci*I	CC↓($^{C}_{G}$)GG[g]
Neisseria denitrificans	*Nde*I	CA↓TATG
	*Nde*II	GATC
Neisseria flavescens	*Nfl*I	GATC
	*Nfl*II	?
	*Nfl*III	?
Neisseria gonorrhoea	*Ngo*I	PuGCGCPy
Nesseria gonorrhoea	*Ngo*II	GGCC
Neisseria gonorrhoea KH 7764-45	*Ngo*III	CCGCGG
Neisseria mucosa	*Nmu*I	GCCGGC
Neisseria ovis	*Nov*I	?
	*Nov*II	GANTC
Nocardia aerocolonigenes	*Nae*I	GCC↓GGC
Nocardia amarae	*Nam*I	GGCGGCC
Nocardia argentinensis	*Nar*I	GG↓CGCC
Nocardia blackwellii	*Nbl*I	CGAT↓CG
Nocardia brasiliensis	*Nbr*I	GCCGGC
Nocardia brasiliensis	*Nba*I	GCCGGC
Nocardia corallina	*Nco*I	C↓CATGG
Nocardia dassonvillei	*Nda*I	GG↓CGCC
Nocardia minima	*Nmi*I	GG↓TACC
Nocardia opaca	*Nop*I	G↓TCGAC
	*Nop*II	?
Nocardia otitidis-caviarum	*Not*I	?
Nocardia otitidis-caviarum	*Noc*I	CTGCAG
Nocardia rubra	*Nru*I	TCG↓CGA

Microorganism	Enzyme	Recognition Sequence
Nocardia uniformis	*Nun*I	?
	*Nun*II	GG↓CGCC
Nostoc (species)	*Nsp*BI	TTCGAA
	*Nsp*BII	$C(^A_C)G{\downarrow}X(^T_G)G$
Nostoc (species)	*Nsp*(7524)I	PuCATG↓Py
	Nsp(7524)II	$G(^A_T)GC(^A_T){\downarrow}C$
	Nsp(7524)III	C↓PyCGPuG
	Nsp(7524)IV	C↓GNCC
	Nsp(7524)V	TTCGAA
Nostoc (species)	*Nsp*HI	PuCATG↓Py
	*Nsp*HII	$GG(^A_T)CC$
Oerskovia xanthineolytica	*Oxa*I	AGCT
	*Oxa*II	?
Proteus vulgaris	*Pvu*I	CGAT↓CG
	*Pvu*II	CAG↓CTG
Providencia alcalifaciens	*Pal*I	GGCC
Providencia stuartii 164	*Pst*I	CTGCA↓G
Pseudoanabaena species	*Psp*I	GGNCC
Pseudomonas aeruginosa	*Pae*R7	?
Pseudomonas facilis	*Pfa*I	GATC
Pseudomonas maltophila	*Pma*I	CTGCAG
Rhizobium leguminosarum 300	*Rle*I	?
Rhizobium lupini #1	*Rlu*I	?
Rhizobrium meliloti	*Rme*I	?
Rhodococcus rhodochrous	*Rrh*I	GTCGAC
	*Rrh*II	?
Rhodococcus rhodochrous	*Rro*I	GTCGAC
Rhodococcus (species)	*Rhs*I	GGATCC
Rhodococcus (Species)	*Rhp*I	GTCGAC
	*Rhp*II	?
Rhodococcus (species)	*Rhe*I	GTCGAC
Rhodospirillum rubrum	*Rrb*I	?
Rhodopseudomonas sphaeroides	*Rsp*I	CGATCG
Rhodopseudomonas sphaeroides	*Rsh*I	CGAT↓CG
Rhodopseudomonas sphaeroides	*Rsa*I	GT↓AC
Rhodopseudomonas sphaeroides	*Rsr*I	GAATTC
Salmonella infantis	*Sin*I	$GG(^A_T)CC$
Serratia marcescens S_b	*Sma*I	CCC↓GGG
Serratia species SAI	*Ssp*I	?
Sphaerotilus natans C	*Sna*I	GTATAC
Spiroplasma citri ASP2	*Sci*NI	G↓CGC
Staphylococcus aureus 3A	*Sau*3A	↓GATC
Staphylococcus aures PS96	*Sau*961	G↓GNCC
Staphyolococcus saprophyticus	*Ssa*I	?

Microorganism	Enzyme	Recognition Sequence
Streptococcus cremoris F	*Scr*FI	CCNGG
Streptococcus durans	*Sdu*I	G$\quad$C G(^A_T)GC(^A_T)C
Streptococcus dysgalactiae	*Sdy*I	GGNCC
Streptococcus faecalis var. *zymogenes*	*Sfa*I	GG↓CC
Streptococcus faecalis GU	*Sfa*GUI	CCGG
Streptococcus faecalis ND547	*Sfa*NI	GCATC(5/9)
Streptomyces achromogenes	*Sac*I	GAGCT↓C
	*Sac*II	CCGC↓GG
	*Sac*III	?
Streptomyces albus	*Sal*PI	CTGCA↓G
Streptomyces albus subspecies *pathocidicus*	*Spa*I	CTCGAG
Streptomyces albus G	*Sal*I	G↓TCGAC
	SalII	?
Streptomyces aureofaciens IKA 18/4	*Sau*I	CC↓TNAGG
Streptomyces bobili	*Sbo*I	CCGCGG
Streptomyces caespitosus	*Sca*I	AGTACT
Streptomyces cupidosporus	*Scu*I	CTCGAG
Streptomyces exfoliatus	*Sex*I	CTCGAG
	*Sex*II	?
Streptomyces fradiae	*Sfr*I	CCGCGG
Streptomyces ganmycicus	*Sga*I	CTCGAG
Streptomyces goshikiensis	*Sgo*I	CTCGAG
Streptomyces griseus	*Sgr*I	?
Streptomyces hygroscopicus	*Shy*TI	?
Streptomyces hygroscopicus	*Shy*I	CCGCGG
Streptomyces lavendulae	*Sla*I	C↓TCGAG
Streptomyces luteoreticuli	*Slu*I	CTCGAG
Streptomyces oderifer	*Sod*I	?
	*Sod*II	?
Streptomyces phaeochromogenes	*Sph*I	GCATG↓C
Streptomyces stanford	*Sst*I	GAGCT↓C
	*Sst*II	CCGC↓GG
	*Sst*III	?
	*Sst*IV	TGATCA
Streptomyces tubercidicus	*Stu*I	AGG↓CCT
Streptoverticillium flavopersicum	*Sfl*I	CTGCA↓G
Thermoplasma acidophilum	*Tha*I	CG↓CG
Thermopolyspora glauca	*Tgl*I	CCGCGG
Thermus aquaticus YTI	*Taq*I	T↓CGÅ*
	*Taq*II	?
Thermus aquaticus	*Taq*XI	C*C↓AGG
Thermus flavus AT62	*Tfl*I	TCGÅ*
Thermus thermophilus HB8	*Tth*HB8I	TCGA
Thermus thermophilus strain 23	*Ttr*I	GACNNGTC
Thermus thermophilus strain 110	*Tte*I	GACNNNGTC

Microorganism	Enzyme	Recognition Sequence
Thermus thermophilus strain 111	*Tth*111I	GACN↓NNGTC
	*Tth*111II	CAAPuCA
	*Tth*111III	?
Tolypothrix tenuis	*Ttn*I	GGCC
Vibrio narveyi	*Vha*I	GGCC
Xanthomonas amaranthicola	*Xam*I	GTCGAC
Xanthomonas badrii	*Xba*I	T↓CTAGA
Xanthomonas holcicola	*Xho*I	C↓TCGAG
	*Xho*II	Pu↓GATCPy
Xanthomonas malvacearum	*Xma*I	C↓CCGGG
	*Xma*II	CTGCAG
	*Xma*III	C↓GGCCG
Xanthomonas manihotis 7 AS1	*Xmn*I	GAANN↓NNTTC
Xanthomonas nigromaculans	*Xni*I	CGATCG
Xanthomonas oryzae	*Xor*I	CTGCAG
	*Xor*II	CGAT↓CG
Xanthomonas papavericola	*Xpa*I	C↓TCGAG

[a] Restriction enzymes are given in alphabetical order. Bases in parentheses show the possible substitution; Pu = purine, Py = pyrimidine, and N = any base. Enzymes such as *Eco*PI and *Eco*P15 are intermediates between type I and II and are known as type III enzymes. (See Table 2.1.) For enzymes such as *Hga*I, (5/10) indicates that cleavage occurs after five nucleotides in the 5′ → 3′ direction and after 10 nucleotides in the 3′ → 5′ direction.

APPENDIX
3

Examples of Isoschizomers[a]

Appendix 3 Examples Of Isoschizomers[a]

AcyI	G p u ↓ C G P y C	CauI	. . . . .
AhaII	. . . ↓	ClmII	. . . . .
AosII	. . . ↓	FdiI	. ↓ . . .
AstWI	. . . ↓	HgiBI	. ↓ . . .
AusIII	. . . ↓	HgiCII	. ↓ . . .
BbiII	. . . .	HgiEI	. ↓ . . .
HgiDI	. . . ↓	HgiHIII	. ↓ . . .
HgiGI	. . . ↓	Nsp7524III	. ↓ . . .
HgiHII	. . . ↓	NspHII	. . . . .
AluI	AG ↓ C T	SinI	. . . . .
OxaI	. .	**BamHI**	G ↓ G A T T C
AsuI	G ↓ G N C C	AacI	. . . . . .
Nsp7524IV	. ↓	AeaI	. . . . . .
PspI	. . . . . .	BamFI	. . . . . .
sau96I	. ↓	BamKI	. . . . . .
SdyI	. . . . . .	BamNI	. . . . . .
AsuII	T T ↓ C G A A	BstI	. ↓ . . . .
MlaI	. . ↓	DdsI	. . . . . .
NspBI	. . . . . .	GdoI	. . . . . .
Nsp7524V	. . . . . .	GoxI	. . . . . .
AvaI	C ↓ P t C G P u G	**BclI**	T ↓ G A T C A
AquI	. . . .	AtuCI	. . . . . .
AvrI	. . . .	BstGI	. . . . . .
	A	CpeI	. . . . . .
AvaII	G ↓ G T C C	SstIV	. . . . . .
AflI	. ↓	**BsePI**	G C G C G C
BamNₓI	. ↓	BssHII	. . . . . .

219

BstEII	G↓GTNACC	*Clt*I	..↓..
*Asp*AI	.↓......	*Fnu*DI	..↓..
*Bst*PI	.↓......	*Hhg*I	..↓..
*Eca*I	.↓......	*Mni*I	....
*Fsp*AI	.↓......	*Mnn*II	....
BstXI	CCA(N)₅↓TGG	*Ngo*II	....
*Bss*GI		*Pal*I	....
*Bst*TI	....	*Sfa*I	..↓.
	C	*Ttn*I	....
CauII	CC↓GGG	*Vha*I	....
*Aha*I	.....	***HgiCI***	G↓GPyPuCC
*Bcn*I	..↓..	*Ban*I	.↓.......
*Nci*I	..↓..	*Hgi*HI	.↓.......
CfrI	(Py)↓GGCC(Pu)	***HgiJII***	GPuGCPy↓C
*Eae*I	.↓.....	*Ban*II	........↓.
DpnI	GA↓TC	*Bvu*I	.........↓.
*Cfu*I	....	***HhaI***	GCG↓C
EcoRI	G↓AATTC	*Cfo*I	....
*Rsr*I	....	*Fnu*DIII	...↓.
	A	*Hin*GUI	....
EcoRII	↓CCTGG	*Hin*P1I	G↓CGC
*Aor*I	..↓...	*Hin*S1	....
*Apy*I	..↓...	*Hin*S2	....
*Atu*BI	.....	*Mnn*IV	....
*Atu*II	.....	*Sci*NI	.↓...
*Bin*SI	.....	***HindII***	GTPy↓PuAC
*Bst*GII	.....	*Chu*II	....
*Bst*NI	..↓..	*Hinc*II	....↓...
*Eca*II	.....	*Hin*JCI	....↓...
*Ecl*II	.....	*Mnn*I	....
*Mph*I	.....	***HindIII***	A↓AGCTT
*Taq*XI	..↓...	*Bbr*I	......
FnuDII	CG↓CG	*Bpe*I	......
*Acc*II	....	*Chu*I	......
*Bce*RI	....	*Hin*173I	......
*Bsu*1192II	....	*Hin*bIII	......
*Hin*1056I	....	*Hin*fII	......
*Tha*I	.↓..	*Hin*JCII	......
FokI	GGATG(9/13)	*Hsu*I	.↓....
*Hin*GUII	.....	*Mki*I	......
HaeII	(Pu)GGCC↓(Py)	***HinfI***	G↓ANTC
*Hin*HI	.....	*Fnu*AI	.↓....
*Ngo*I	.....	*Hha*II	....
HaeIII	GG↓CC	*Nca*I	....
*Blu*II	....	*Nov*II	....
*Bse*I	....	***HpaI***	GTT↓AAC
*Bsp*RI	..↓.	*Bse*II	....
*Bss*CI	....	***HpaII***	C↓CGG
*Bst*CI	....	*Bsu*1192I	....
*Bsu*1076I	....	*Bsu*1231I	....
*Bsu*1114I	....	*Hap*II	.↓...
*Bsu*RI	..↓..	*Mni*II	....
*Clm*I	....	*Mno*I	.↓...
MspI	.↓....	*Rsp*I	......

SfaGUI	. . . .	*Xni*I	. . . . . .
KpnI	G G T A C ↓ C	*Xor*II	. . . . ↓ . .
*Nmi*I	. . . . . .	**Sac**I	G A G C T ↓ C
MboI	↓ G A T C	*Sst*I	. . . . . ↓ .
*Bsa*PI	. . . .	**Sac**II	C C G C ↓ G G
*Bsr*PII	. . . .	*Bac*I	. . : . . .
*Bss*GII	. . . .	*Csc*I	. . . ↓ . .
*Bss*EIII	. . . .	*Ecc*I	. . . . . .
*Cpa*I	. . . .	*Mra*I	. . . . . .
*Dpn*II	. . . .	*Ngi*III	. . . . . .
*Fnu*AII	. . . .	*Sbo*I	. . . . . .
*Fnu*CI	↓	*Sfr*I	. . . . . .
*Fnu*EI	↓	*Shy*I	. . . . . .
*Mno*III	. . . .	*Sst*II	. . . ↓ . .
*Mos*I	. . . .	*Tgl*I	. . . . . .
*Mth*I	. . . .	**Sal**I	G ↓ T C G A C
*Nde*II	. . . .	*Hgi*CIII	. ↓ . . . .
*Nfl*I	. . . .	*Hgi*DII	. ↓ . . . .
MstI	T G C ↓ G C A	*Nop*I	. ↓ . . . .
*Aos*I	. . . ↓ . . .	*Rhe*I	. . . . . .
*Fdi*II	. . . ↓ . . .	*Rhp*I	. . . . . .
MstII	C C ↓ T N A G G	*Rrh*I	. . . . . .
*Cvn*I	. . ↓	*Rro*I	. . . . . .
*Sau*I	. . ↓	*Xam*I	. . . . . .
NaeI	G C C ↓ G G C	**Stu**I	A G G ↓ C C T
*Nba*I	. . . ↓ . . .	*Aat*I	. . . . . .
*Nbr*I	. . . ↓ . . .	*Gdi*I	. . . ↓ . . .
*Nmu*I	. . . . . .	**Taq**I	T ↓ C G A
NarI	G G ↓ C G C C	*Tfi*I	. . . .
*Bbe*AI	. . . . . .	*Tth*HB8	. . . .
*Bde*I	G G C G C ↓ C	**Tth**111I	G A C N ↓ N N G T C
*Bin*SII	. . . . .	*Tte*I	. . . .
*Nam*I	. . . . . .	*Ttr*I	. . . .
*Nda*I	. . . . . .	**Xho**I	C ↓ T C G A G
*Nun*II	. . ↓ . . .	*Bbi*III	. . . . . .
NurI	T C G ↓ C G A	*Blu*I	. ↓ . . . .
*Ama*I	. . . . . .	*Bss*HI	. . . . . .
NspCI	P u C A T G ↓ P y	*Bst*HI	. . . . . .
*Nsp*HI	. . . . . . ↓ . .	*Bth*I	. . . . . .
PstI	C T G C A ↓ G	*Ccr*II	. . . . . .
*Bbi*I	. . . . . .	*Dde*II	. . . . . .
*Bce*170I	. . . . . .	*Msi*I	. . . . . .
*Bsu*1247I	. . . . . .	*Pae*R7I	. ↓ . . . .
*Noc*I	. . . . . .	*Scu*I	. . . . . .
*Pma*I	. . . . . .	*Sex*I	. . . . . .
*Sal*PI	. . . . . ↓ .	*Sga*I	. . . . . .
*Sfi*I	. . . . . ↓ .	*Sgo*I	. . . . . .
*Xma*II	. . . . . .	*Sla*I	. ↓ . . . .
*Xor*I	. . . . . .	*Slu*I	. . . . . .
Pvu*I*	C G A T ↓ C G	*Spa*I	. . . . . .
*Nbl*I	. . . . ↓ . .	*Xpa*I	. . . . . .
*Rsh*I	. . . . ↓ . .		

---

[a] The first enzyme in each isochizomer series is shown in bold; the others are abbreviated. Arrows show the cleavage sites when known.

APPENDIX

4

Methylation Chart (From McClelland, 1983)[a]

Appendix 4 Methylation Chart (From McClelland, 1983)[a]

Restriction Enzyme	Recognition Site	Methylated Sequence Cut	Methylated Sequence Uncut	Methylation Effect Unknown
*Alu*I	AGCT	?	AG^mCT	^mAGCT
*Hap*II	CCGG	?	C^mCGG ≠	^mCCGG
*Hpa*II	CCGG	^mCCGG	C^mCGG ≠	—
*Msp*I	CCGG	C^mCGG	^mCCGG ≠	—
*Bst*EIII	GATC	?	G^mATC	GAT^mC
*Dpn*I	GATC	G^mATC	GATC	Only cut methylated DNA
*Dpn*II	GATC	?	G^mATC*	GAT^mC
*Fnu*AII	GATC	?	G^mATC	GAT^mC
*Fnu*CI	GATC	?	G^mATC	GAT^mC
*Fnu*EI	GATC	G^mATC	?	GAT^mC*
*Mbo*I	GATC	?	G^mATC	GAT^mC
*Mno*III	GATC	?	G^mATC	GAT^mC
*Mos*I	GATC	?	G^mATC	GAT^mC
*Mph*I	GATC	?	G^mATC	GAT^mC
*Pfa*I	GATC	G^mATC	?	GAT^mC*
*Sau*3A	GATC	G^mATC	GAT^mC*	—
*Hha*I	GCGC	—	G^mCGC ≠ GCG^mC	—
*Bsu*RI	GGCC	?	GG^mCC ≠	GGC^mC
*Hae*III	GGCC	GGC^mC	GG^mCC ≠	—
*Taq*I	TCGA	T^mCGA	TCG^mA ≠	—
*Tth*I	TCGA	T^mCGA	TCG^mA ≠	—

Restriction Enzyme	Recognition Site	Methylated Sequence Cut	Methylated Sequence Uncut	Methylation Effect Unknown
*Hin*fI	GANTC	GANT^{m}C	?	G^{m}ANTC*
*Fnu*4HI	GCNGC	?	G^{m}CNGG	GCNG^{m}C
*Sau*96	GGNCC	?	GGNC^{m}C	GGN^{m}CC
*Aac*I	CCXGG	C^{m}CXGG	?	^{m}CCXGG*
*Apy*I	CCXGG	C^{m}CXGG	^{m}CCXGG*	—
*Atu*BI	CCXGG	?	C^{m}CXGG	^{m}CCXGG
*Atu*II	CCXGG	?	C^{m}CXGG	^{m}CCXGG
*Bst*NI	CCXGG	C^{m}CXGG MCCXGG(b)	?	?
*Eca*II	CCXGG	?	C^{m}CXGG	^{m}CCXGG
*Ecl*II	CCXGG	?	C^{m}CXGG	^{m}CCXGG
*Eco*RII	CCXGG	^{m}CCXGG	C^{m}CXGG ≠	—
*Mph*I	CCXGG	?	C^{m}CXGG	^{m}CCXGG
*Taq*XI	CCXGG	^{m}CCXGG	?	C^{m}CXGG
*Nci*I	CCZGG	?	C^{m}CZGG	^{m}CCZGG
*Bbv*I	GCXGC	?	G^{m}CXGC ≠	GCXG^{m}C
*Ava*II	GGXCC	?	GGXC^{m}C	GGX^{m}CC
*Eco*PI	AGACC	?	AG^{m}ACC ≠ AGA^{m}CC AGAC^{m}C GGT^{m}CT	^{m}AGACC
*Mbo*II	GAAGA	?	GAAG^{m}A Cmethylation	G^{m}AAGA GA^{m}AGA
*Ava*I	CYCGRG	?	CY^{m}CGRY	^{m}CYCGRG
*Aos*II	GRCGYC	?	GR^{m}CGYC	GRCGY^{m}C
*Acc*I	GTJKAC	?	GTJK^{m}AC	GTJKA^{m}C
*Hin*dII	GTYRAC	GTYRA^{m}C	GTYR^{m}AC ≠	—
*Hae*II	RGCGCY	?	RG^{m}CGCY	RGCG^{m}CY
*Xho*II	RGATCY	RG^{m}ATCY	RGAT^{m}CY*	
*Hin*dIII	AAGCTT	?	^{m}AAGCTT ≠ AAG^{m}CTT	A^{m}AGCTT
*Bgl*II	AGATCT	AG^{m}ATCT	AGAT^{m}CT	^{m}AGATCT
*Cla*I	ATCGAT	?	ATCG^{m}AT ≠	AT^{m}CGAT ^{m}ATCGAT
*Pvu*II	CAGCTG	?	C methylation	?
*Sma*I	CCCGGG	?	CC^{m}CGGG	C^{m}CCGGG ^{m}CCCGGG
*Xma*I	CCCGGG	?	CC^{m}CGGG C^{m}CCGGG	^{m}CCCGGG
*Sac*II	CCGCGG	?	C methylation	?
*Pvu*I	CGATCG	CG^{m}ATCG	CGAT^{m}CG	^{m}CGATCG
*Xor*II	CGATCG	CG^{m}ATCG	CGAT^{m}CG	^{m}CGATCG
*Xma*III	CGGCCG	?	CGG^{m}CCG	^{m}CGGCCG CGGC^{m}CG
	CTCGAC	?	CT^{m}CGAG CTCG^{m}AG	^{m}CTCGAG

Restriction Enzyme	Recognition Site	Methylated Sequence Cut	Methylated Sequence Uncut	Methylation Effect Unknown
*Pst*I	CTGCAG	?	C methylation	?
*Sal*PI	CTGCAG	?	C methylation	?
*Eco*RI	GAATTC	?	GA^{m}ATTC $\neq$ GAATT^{m}C	G^{m}AATTC
*Bam*HI	GGATCC	GGATC^{m}C GG^{m}ATCC	GGAT^{m}CC $\neq$	—
*Apa*I	GGGCCC	?	GGG^{m}CCC	GGGC^{m}CC GGGCC^{m}C
*Sal*I	GTCGAC	?	GTCG^{m}AC GT^{m}CGAC	GTCGA^{m}C
*Hpa*I	GTTAAC	GTTAA^{m}C	GTTA^{m}AC $\neq$	GTT^{m}AAC
*Xba*I	TCTAGA	?	T^{m}CTAGA	TCT^{m}AGA TCTAG^{m}A
*Atu*CI	TGATCA	?	TG^{m}ATCA	TGAT^{m}CA TGATC^{m}A
*Bcl*I	TGATCA	TGAT^{m}CA	TG^{m}ATCA	TGATC^{m}A
*Cpe*I	TGATCA	?	TG^{m}ATCA	TGAT^{m}CA TGATC$_m$A
*Bal*I	TGGCCA	?	TGG^{m}CCA	TGGC^{m}CA TGGCC^{m}A

[a] The symbol # indicates that a methylase is known; *, that its existence is likely. The letters represent the following nucleotides: J = A or C; K = G or T; N = A, G, C or T; R = A or G; Y = C or T; X = A or T; Z = G or C.

5

Detailed Genetic and Physical Map of λ DNA (From Williams and Blattner, 1980)

Functionally related groups of genes are bracketed, the genetic and restriction maps are shown below.

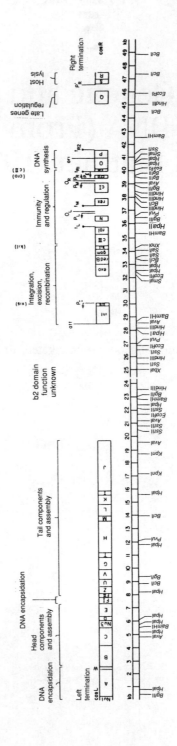

6

Simplified Restriction Map of λ DNA (From Rodriguez and Tait, 1983)

The map is given in kbp. Numbers indicate the distance of each restriction site from the left end of the phage genome. For a more detailed map see Szybalski and Szybalski (1979).

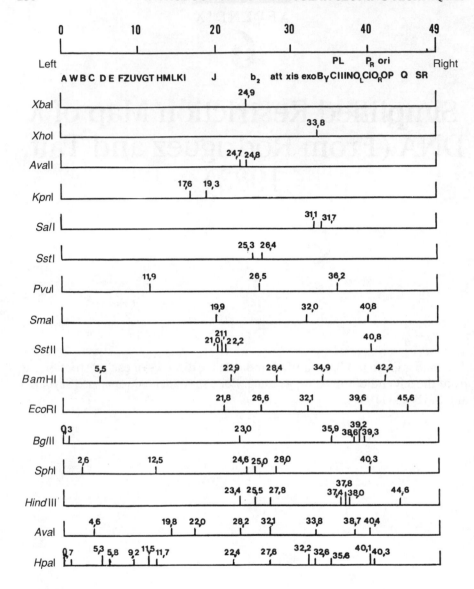

7

Map of the First 21 Vectors of the Charon Series (From De Wett et al., 1980 and Rimm et al., 1980)

*lac*5, *bio*1, and *bio*256 are substitutions of the *lac* and *bio* regions of *E. coli;* *att*[80], *imm*[80], and *QSR*[80] are substitutions of ϕ80; *b*1007 is a deletion in the *b* region that affects the *att* site and prevents lysogeny.

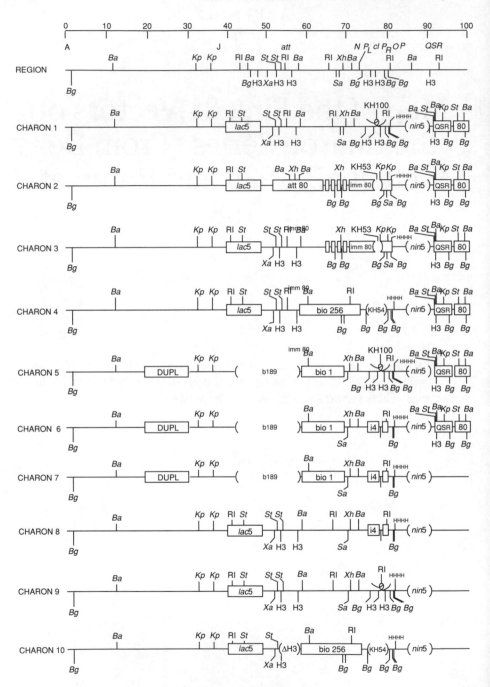

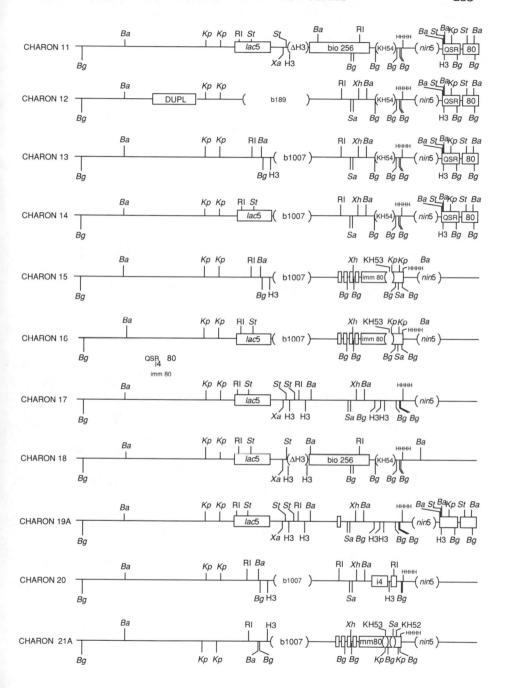

Complete Nucleotide Sequence of pBR322 (From Different Authors Quoted in the Text)

The map starts at the *Eco*RI site of pBR322 and the additional base doublet is indicated by two vertical arrowheads (4,363 bp). The promoter regions of the tetracyclin resistance and β-lactamase genes are shown by wavy arrows. The origin of replication is labeled with a bold arrow. The positions of a few restriction sites are also shown in bold characters. Start and stop codons are boxed.

*Hind*III

EcoRI   *Tc* promoter
(GAA)TTCTCATGTTTGACAGCTTATCATCGATAAGCTTIAATGCGGTAGTTTATCACAGTTAAATTGCTAACGCAGTCAGGCACCGTGTATGAAATCTAACAAT
                                                                                                        100.
AAGAGTACAAACTGTCGAATAGTAGCTATTCGAAATTACGCCATCAAATAGTGTCAATTTAACGATTGCGTCAGTCCGTGGCACATACTTTAGATTGTTA
    a-*Tc* promoter

GCGCTCATCGTCATCCTCGGCACCGTCACCCTGGATGCTGTAGGCATAGGCTTGGTTATGCCGGTACTGCCGGGCCTCTTGCGGGGATATCGTCCATTCCG
                                                                                                        200.
CGCGAGTAGCAGTAGGAGCCGTGGCAGTGGGACCTACGACATCCGTATCCGAACCAATACGGCCATGACGGCCCGGAGAACGCCCTATAGCAGGTAAGGC

ACAGCATCGCCAGTCACTATGGCGTGCTGCTAGCGCTATATGCGTTGATGCAATTTCTATGCGCACCCGTTCTCGGAGCACTGTCCGACCGCTTTGGCCG
                                                                                                        300.
TGTCGTAGCGGTCAGTGATACCGCACGACGATCGCGATATACGCAACTACGTTAAAGATACGCGTGGGCAAGAGCCTCGTGACAGGCTGGCGAAACCGGC

                                                                          *Bam*HI
CCGCCCAGTCCTGCTCGCTTCGCTACTTGGAGCCACTATCGACTACGCGATCATGGCGACCACACCCGTCCTGTGTGGATCCTCTACGCCGGACGCATCGTG
                                                                                                        400.
GGCGGGTCAGGACGAGCGAAGCGATGAACCTCGGTGATAGCTGATGCGCTAGTACCGCTGGTGTGGGCAGGACACCTAGGAGATGCGGCCTGCGTAGCAC

GCCGGCATCACCGGCGCCACAGGTGCGGTTGCTGGCGCCTATATCGCCGACATCACCGATGGGGAAGATCGGGCTCGCCACTTCGGGCTCATGAGCGCTT
                                                                                                        500.
CGGCCGTAGTGGCCGCGGTGTCCACGCCAACGACCGCGGATATAGCGGCTGTAGTGGCTACCCCTTCTAGCCCGAGCGGTGAAGCCCGAGTACTCGCGAA

                                                      *Sph*I
GTTTCGGCGTGGGTATGGTGGCAGGCCCGTGGCCGGGGGACTGTTGGGCGGCCATCTCCTTGCATGCACCATTCCTTGCGGCGGCGGTGCTCAACGGCCTC
                                                                                                        600.
CAAAGCCGCACCCATACCACCGTCCGGGCACCGGCCCCCTGACAACCCGCGGTAGAGGAACGTACGTGGTAAGGAACGCCGCCGCCACGAGTTGCCGGAG

                                                  *Sal*I
AACCTACTACTGGGCTGCTTCCTAATGCAGGAGTCGCATAAGGGAGAGCGTLGACCGATGCCCTTGAGAGCCTTCAACCCAGTCAGCTCCTTCCGGTGGG
                                                                                                        700.
TTGGATGATGACCCGACGAAGGATTACGTCCTCAGCGTATTCCCTCTCGCAGCTGGCTACGGGAACTCTCGGAAGTTGGGTCAGTCGAGGAAGGCCACCC

CGCGGGGCATGACTATCGTCGCCGCACTTATGACTGTCTTCTTTATCATGCAACTCGTAGGACAGGTGCCGGCAGCGCTCTGGGTCATTTTCGGCGAGGA
                                                                                                        800.
GCGCCCCGTACTGATAGCAGCGGCGTGAATACTGACAGAAGAAATAGTACGTTGAGCATCCTGTCCACGGCCGTCGCGAGACCCAGTAAAAGCCGCTCCT

CCGCTTTCGCTGGAGCGCGACGATGATCGGCCTGTCGCTTGCGGTATTCGGAATCTTGCACGCCCTCGCTCAAGCCTTCGTCACTGGTCCCGCCACCAAA
                                                                                                        900.
GGCGAAAGCGACCTCGCGCTGCTACTAGCCGGACAGCGAACGCCATAAGCCTTAGAACGTGCGGGAGCGAGTTCGGAAGCAGTGACCAGGGCGGTGGTTT

CGTTTCGGCGAGAAGCAGGCCATTATCGCCGGCATGGCGGCCGACGCGCTGGGCTACGTCTTGCTGGCGTTCGCGACGCGAGGCTGGATGGCCTTCCCCA
                                                                                                       1000.
GCAAAGCCGCTCTTCGTCCGGTAATAGCGGCCGTACCGCCGGCTGCGCGACCCGATGCAGAACGACCGCAAGCGCTGCGCTCCGACCTACCGGAAGGGGT

TTATGATTCTTCTCGCTTCCGGCGGCATCGGGATGCCCGCGTTGCAGGCCATGCTGTCCAGGCAGGTAGATGACGACCATCAGGGACAGCTTCAAGGATC
                                                                                                       1100.
AATACTAAGAAGAGCGAAGGCCGCCGTAGCCCTACGGGCGCAACGTCCGTACGACAGGTCCGTCCATCTACTGCTGGTAGTCCCTGTCGAAGTTCCTAG

GCTCGCGGCTCTTACCAGCCTAACTTCGATCACTGGACCGCTGATCGTCACGGCGATTTATGCCGCCTCGGCGAGCACATGGAACGGGTTGGCATGGATT
                                                                                                       1200.
CGAGCGCCGAGAATGGTCGGATTGAAGCTAGTGACCTGGCGACTAGCAGTGCCGCTAAATACGGCGGAGCCGCTCGTGTACCTTGCCCAACCGTACCTAA

GTAGGCGCCGCCCTATACCTTGTCTGCCTCCCCGCGTTGCGTCGCGGTGCATGGAGCCGGGCCACCTCGACCTGAATGGAAGCCGGCGGCACCTCGCTAA
                                                                                                       1300.
CATCCGCGGCGGGATATGGAACAGACGGAGGGGCGCAACGCAGCGCCACGTACCTCGGCCCGGTGGAGCTGGACTTACCTTCGGCCGCCGTGGAGCGATT

CGGATTCACCACTCCAAGAATTGGAGCCAATCAATTCTTGCGGAGAACTGTGAATGCGCAAACCAACCCTTGGCAGAACATATCCATCGCGTCCGCCATC
                                                                                                       1400.
GCCTAAGTGGTGAGGTTCTTAACCTCGGTTAGTTAAGAACGCCTCTTGACACTTACGCGTTTGGTTGGGAACCGTCTTGTATAGGTAGCGCAGGCGGTAG

                              *Ava*I   *Bal*I
TCCAGCAGCCGCACGCGGCGCATCTCGGCAGCGTTGGGTCCTCGCCACGGGTGCGCATGATCGTGCTCCTGTCGTTGAGGACCCGGCTAGGCTGGCGGG
                                                                                                       1500.
AGGTCGTCGGCGTGCGCCGCGTAGAGCCCGTCGCAACCCAGGACCGGTGCCCACGCGTACTAGCACGAGGACAGCAACTCCTGGGCCGATCGGACCGCCC

GTTGCCTTACTGGTTAGCAGAATGAATCACCGATACGCGAGCGAACGTGAAGCGACTGCTGCTGCAAAACGTCTGCGACCTGAGCAACAACATGAATGGT
                                                                                                       1600.
CAACGGAATGACCAATCGTCTTACTTAGTGGCTATGCGCTCGCTTGCACTTCGCTGACGACGACGTTTTGCAGACGCTGGACTCGTTGTTGTACTTACCA

CTTCGGTTTCCGTGTTTCGTAAAGTCTGGAAACGCGGAAGTCAGCGCCCTGCACCATTATGTTCCGGATCTGCATCGCAGGATGCTGCTGGCTACCCTGT
                                                                                                       1700.
GAAGCCAAAGGCACAAAGCATTTCAGACCTTTGCGCCTTCAGTCGCGGGACGTGGTAATACAAGGCCTAGACGTAGCGTCCTACGACGACCGATGGGACA

GGAACACCTACATCTGTATTAACGAAGCGCTGGCATTGACCCTGAGTGATTTTTCTCTGGTCCCGCCGCATCCATACCGGCCAGTTGTTTACCCTCACAAC
                                                                                                       1800.
CCTTGTGGATGTAGACATAATTGCTTCGCGACCGTAACTGGGACTCACTAAAAAGACACCAGGGCGGCGTAGGTATGGCGGTCAACAAATGGGAGTGTTG

GTTCCAGTAACCGGGCATGTTCATCATCAGTAACCCGTATCGTGAGCATCCTCTCTCGTTTCATCGGTATCATTACCCCCATGAACAGAAATTCCCCCTT
                                                                                                       1900.
CAAGGTCATTGGCCCGTACAAGTAGTAGTCATTGGGCATAGCACTCG*AGGAGAGAGCAAAGTAGCCATAGTAATGGGGGTACTTGTCTTTAAGGGGGAA

ACACGGAGGCATCAAGTGACCAAACAGGAAAAAACCGCCCTTAACATGGCCCGCTTTATCAGAAGCCAGACATTAACGCTTCTGGAGAAACTCAACGAGC
                                                                                                       2000.
TGTGCCTCCGTAGTTCACTGGTTTGTCCTTTTTTTGGCGGGAATTGTACCGGGCGAAATAGTCTTCGGTCTGTAATTGCGAAGACCTCTTTGAGTTGCTCG

Pvu II

```
              TGGACGCGGATGAACAGGCAGACATCTGTGAATCGCTTCACGACCACGCTGATGAGCTTTACCGCAGCTGCCTCGCGCGTTTCGGTGATGACGGTGAAAA
                                                                                                    2100.
              ACCTGCGCCTACTTGTCCGTCTGTAGACACTTAGCGAAGTGCTGGTGCGACTACTCGAAATGGCGTCGACGGAGCGCGCAAAGCCACTACTGCCACTTTT

              CCTCTGACACATGCAGCTCCCGGAGACGGTCACAGCTTGTCTGTAAGCGGATGCCGGGAGCAGACAAGCCCGTCAGGGCGCGTCAGCGGGTGTTGGCGGG
                                                                                                    2200.
              GGAGACTGTGTACGTCGAGGGCCTCGCCAGTGTCGAACAGACATTCGCCTACGGCCCTCGTCTGTTCGGGCAGTCCCGCGCAGTCGCCCACAACCGCCC

              TGTCGGGGGCGAGCCATGACCCAGTCACGTAGCGATAGCGGAGTGTATACTGGCTTAACTATGCGGCATCAGAGCAGATTGTACTGAGAGTGCACCATAT
                                                                                                    2300.
              ACAGCCCCGCGTCGGTACTGGGTCAGTGCATCGCTATCGCCTCACATATGACCGAATTGATACGCCGTAGTCTCGTCTAACATGACTCTCACGTGGTATA

              GCGGTGTGAAATACCGCACAGATGCGTAAGGAGAAAATACCGCATCAGGCGCTCTTCCGCTTCCTCGCTCACTGACTCGCTGCGCTCGGTCGTTCGGCTG
                                                                                                    2400.
              CGCCACACTTTATGGCGTGTCTACGCATTCCTCTTTTATGGCGTAGTCCGCGAGAAGGCGAAGGAGCGAGTGACTGAGCGACGCGAGCCAGCAAGCCGAC

              CGGCGAGCGGTATCAGCTCACTCAAAGGCGGTAATACGGTTATCCACAGAATCAGGGGATAACGCAGGAAAGAACATGTGAGCAAAAGGCCAGCAAAAGG
                                                                                                    2500.
              GCCGCTCGCCATAGTCGAGTGAGTTTCCGCCATTATGCCAATAGGTGTCTTAGTCCCCTATTGCGTCCTTTCTTGTACACTCGTTTTCCGGTCGTTTTCC
```

▼ *ori*

```
              CCAGGAACCGTAAAAAGGCCGCGTTGCTGGCGTTTTTCCATAGGCTCCGCCCCCCTGACGAGCATCACAAAAATCGACGCTCAAGTCAGAGGTGGCGAAA
                                                                                                    2600.
              GGTCCTTGGCATTTTTCCGGCGCAACGACCGCAAAAAGGTATCCGAGGCGGGGGGACTGCTCGTAGTGTTTTTAGCTGCGAGTTCAGTCTCCACCGCTTT

              CCCGACAGGACTATAAAGATACCAGGCGTTTCCCCCTGGAAGCTCCCTCGTGCGCTCTCCTGTTCCGACCCTGCCGCTTACCGGATACCTGTCCGCCTTT
                                                                                                    2700.
              GGGCTGTCCTGATATTTCTATGGTCCGCAAAGGGGGACCTTCGAGGGAGCACGCGAGAGGACAAGGCTGGGACGGCGAATGGCCTATGGACAGGCGGAAA

              CTCCCTTCGGGAAGCGTGGCGCTTTCTCAATGCTCACGCTGTAGGTATCTCAGTTCGGTGTAGGTCGTTCGCTCCAAGCTGGGCTGTGTGCACGAACCCC
                                                                                                    2800.
              GAGGGAAGCCCTTCGCACCGCGAAAGAGTTACGAGTGCGACATCCATAGAGTCAAGCCACATCCAGCAAGCGAGGTTCGACCCGACACGTGCTTGGGG

              CCGTTCAGCCCGACCGCTGCGCCTTATCCGGTAACTATCGTCTTGAGTCCAACCCGGTAAGACACGACTTATCGCCACTGGCAGCAGCCACTGGTAACAG
                                                                                                    2900.
              GGCAAGTCGGGCTGGCGACGCGGAATAGGCCATTGATAGCAGAACTCAGGTTGGGCCATTCTGTGCTGAATAGCGGTGACCGTCGTCGGTGACCATTGTC
```

104 bases RNA ∿∿∿∿∿

```
              GATTAGCAGAGCGAGGTATGTAGGCGGTGCTACAGAGTTCTTGAAGTGGTGGCCTAACTACGGCTACACTAGAAGGACAGTATTTGGTATCTGCGCTCTG
                                                                                                    3000.
              CTAATCGTCTCGCTCCATACATCCGCCACGATGTCTCAAGAACTTCACCACCGGATTGATGCCGATGTGATCTTCCTGTCATAAACCATAGACGCGAGAC
```

∿∿∿∿∿

```
              CTGAAGCCAGTTACCTTCGGGAAAAAGAGTTGGTAGCTCTTGATCCGGCAAACAAACCACCGCTGGTAGCGGTGGTTTTTTTGTTTGCAAGCAGCAGATTA ➔
                                                                                                    3100.
              GACTTCGGTCAATGGAAGCCCTTTTTCTCAACCATCGAGAACTAGGCCGTTTGTTTGGTGGCGACCATCGCCACCAAAAAAACAAACGTTCGTCGTCTAAT

              CGCGCAGAAAAAAAGGATCTCAAGAAGATCCTTTGATCTTTTCTACGGGGTCTGACGCTCAGTGGAACGAAAACTCACGTTAAGGGATTTTGGTCATGAG
                                                                                                    3200.
              GCGCGTCTTTTTTTCCTAGAGTTCTTCTAGGAAACTAGAAAAGATGCCCCAGACTGCGAGTCACCTTGCTTTTGAGTGCAATTCCCTAAAACCAGTACTC

              ATTATCAAAAAGGATCTTCACCTAGATCCTTTTAAATTAAAAATGAAGTTTTAAATCAATCTAAAGTATATATGAGTAAACTTGGTCTGACAGTTACCAA
                                                                                                    3300.
              TAATAGTTTTTCCTAGAAGTGGATCTAGGAAAATTTAATTTTTACTTCAAAATTTAGTTAGATTTCATATATACTCATTTGAACCAGACTGTC A̲T̲GGTT

              TGCTTAATCAGTGAGGCACCTATCTCAGCGATCTGTCTATTTCGTTCATCCATAGTTGCCTGACTCCCCGTCGTGTAGATAACTACGATACGGGAGGGCT
                                                                                                    3400.
              ACGAATTAGTCACTCCGTGGATAGAGTCGCTAGACAGATAAAGCAAGTAGGTATCAACGGACTGAGGGGCAGCACATCTATTGATGCTATGCCCTCCCGA

              TACCATCTGGCCCCAGTGCTGCAATGATACCGCGAGACCCACGCTCACCGGCTCCAGATTTATCAGCAATAAACCAGCCAGCCGGAAGGGCCGAGCGCAG
                                                                                                    3500.
              ATGGTAGACCGGGGTCACGACGTTACTATGGCGCTCTGGGTGCGAGTGGCCGAGGTCTAAATAGTCGTTATTTGGTCGGTCGGCCTTCCCGGCTCGCGTC

              AAGTGGTCCTGCAACTTTATCCGCCTCCATCCAGTCTATTAATTGTTGCCGGGAAGCTAGAGTAAGTAGTTCGCCAGTTAATAGTTTGCGCAACGTTGTT
                                                                                                    3600.
              TTCACCAGGACGTTGAAATAGGCGGAGGTAGGTCAGATAATTAACAACGGCCCTTCGATCTCATTCATCAAGCGGTCAATTATCAAACGCGTTGCAACAA
```

*Pst*I

```
              GCCATTGCTGCAGGCATCGTGGTGTCACGCTCGTCGTTTGGTATGGCTTCATTCAGCTCCGGTTCCCAACGATCAAGGCGAGTTACATGATCCCCCATGT
                                                                                                    3700.
              CGGTAACGACGTCCGTAGCACCACAGTGCGAGCAGCAAACCATACCGAAGTAAGTCGAGGCCAAGGGTTGCTAGTTCCGCTCAATGTACTAGGGGGTACA
```

*Pvu*I

```
              TGTGCAAAAAAGCGGTTAGCTCCTTCGGTCCTCCGATCGTTGTCAGAAGTAAGTTGGCCGCAGTGTTATCACTCATGGTTATGGCAGCACTGCATAATTC
                                                                                                    3800.
              ACACGTTTTTTCGCCAATCGAGGAAGCCAGGAGGCTAGCAACAGTCTTCATTCAACCGGCGTCACAATAGTGAGTACCAATACCGTCGTGACGTATTAAG

              TCTTACTGTCATGCCATCCGTAAGATGCTTTTCTGTGACTGGTGAGTACTCAACCAAGTCATTCTGAGAATAGTGTATGCGGCGACCGAGTTGCTCTTGC
                                                                                                    3900.
              AGAATGACAGTACGGTAGGCATTCTACGAAAAGACACTGACCACTCATGAGTTGGTTCAGTAAGACTCTTATCACATACGCCGCTGGCTCAACGAGAACG

              CCGGCCGTCAACACGGGATAATACCGCGCCACATAGCAGAACTTTAAAAGTGCTCATCATTGGAAAACGTTCTTCGGGGCGAAAACTCTCAAGGATC1TAC
                                                                                                    4000.
              GGCCGGCAGTTGTGCCCTATTATGGCGCGGTGTATCGTCTTGAAATTTTCACGAGTAGTAACCTTTTGCAAGAAGCCCCGCTTTTGAGAGTTCCTAGAATG

              CGCTGTTGAGATCCAGTTCGATGTAACCCACTCGTGCACCCAACTGATCTTCAGCATCTTTTACTTTCACCAGCGTTTCTGGGTGAGCAAAAACAGGAAG
                                                                                                    4100.
              GCGACAACTCTAGGTCAAGCTACATTGGGTGAGCACGTGGGTTGACTAGAAGTCGTAGAAATGAAAGTGGTCGCAAAGACCCACTCGTTTTTGTCCTTC

              GCAAAATGCCGCAAAAAAGGGAATAAGGGCGACACGGAAATGTTGAATACTCATCACTCTTCCTTTTTCAATATTATTGAAGCATTTATCAGGGTTATTGT
                                                                                                    4200.
              CGTTTTACGGCGTTTTTTCCCTTATTCCCGCTGTGCCTTTACAACTTATGASTATGAGAAGGAAAAAGTTATAATAACTTCGTAAATAGTCCCAATAACA
```

◀━━ *Ap'* promoter

```
              CTCATGAGCCGGATACATATTTGAATGTATTTAGAAAAATAAACAAATAGGGGTTCCGCGCACATTTCCCCGAAAAGTGCCACCTGACGTCTAAGAAACCA
                                                                                                    4300.
              GAGTACTCGCCTATGTATAAACTTACATAAATCTTTTTATTTGTTTATCCCCAAGGCGCGTGTAAAGGGGCTTTTCACGGTGGACTGCAGATTCTTTGGT
```

*Eco*RI

```
              TTATTATCATGACATTAACCTATAAAAATAGGCGTATCACGAGGCCCTTTCGTCTTCAAGA(TTC)
              AATAATAGTACTGTAATTGGATATTTTTTATCCGCATAGTGCTCCGGGAAAGCAGAAGTTCTT
```

APPENDIX

9

Detailed Restriction Map of pBR322 (From Different Authors)

The central circle gives the position of the unique restriction sites and the scale in 1,000-bp segments. Concentric circles show the positions of different multiple restriction sites and the names of the corresponding fragments.

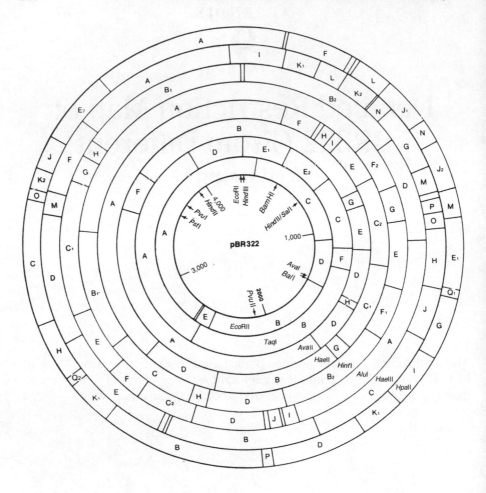

10

Structure of the pUC Plasmid Series

Vectors from the pUC series were constructed using a 2,297-bp *Eco*RI-*Pvu*II fragment from pBR322 (which contains the origin of replication [*ori*] and the gene for β-lactamase) and a 433-bp *Hae*II fragment coding for the promoter (P), operator (O), and α-peptide of the *E. coli* β-galactosidase (*lacZ*) gene. The restriction sites in the plasmid and the multiple-cloning-site regions from pUC7, 8, and 9 are shown.

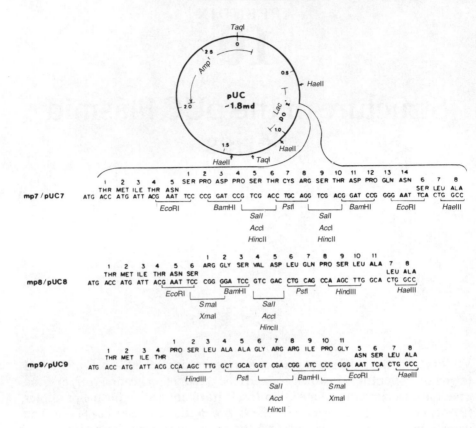

11

Proteinase K

Proteinase K is a very active, nonspecific, protease extracted from the mushroom *Tritirachium album limber*. It is commercially available in a nuclease-free form and is active in a wide range of laboratory conditions (7.4 < pH < 11.5, unaffected by EDTA or metal cations). In addition, the enzyme remains active at temperatures of up to 65 °C and is not deactivated in the presence of 0.1% SDS. In fact, there is a synergistic effect of SDS on proteinase K activity, probably due to the fact that SDS denatures the substrates of the enzyme and thus facilitates its action.

12

Extraction of Lymphocyte DNA

Recover 20 ml of blood and adjust the EDTA concentration to 5%. Centrifuge the sample at 1,500 rpm for 5 min and remove the plasma with a Pasteur pipette connected to a vacuum pump, being careful to leave the white-cell zone intact. Lyse the remaining red blood cells in a hypotonic solution of Tris-HCl, pH 7.6; 10 mM NaCl; 5 mM EDTA. Spin at 2,000 rpm for 10 min, resuspend the pellet in a white-blood-cell lysis solution containing SDS and proteinase K, and incubate overnight at 42 °C. Extract the proteins with several rounds of phenol–chloroform–isoamyl alcohol extraction, followed by several rounds of chloroform–isoamyl alcohol extraction. Recover the aqueous phase and precipitate the DNA with two volumes of ethanol and 0.1 volume of 3 M sodium acetate, pH 5.4, at −20 °C overnight. Rinse the pelleted material in 70% ethanol, dry, and resuspend the DNA in TE with rotary agitation for several days at 4 °C with a few exposures to room temperature. About 2 mg of DNA is obtained by using this method.

13

RNase A

Ribonuclease A (RNase A), extracted from beef pancreas, is an endoribonuclease that specifically attacks the pyrimidines in ribonucleotides and cuts the 5′ phosphate bond of the adjoining molecule.

14

DNase I

Deoxyribonuclease I (DNase I), also extracted from beef pancreas, is an endonuclease that hydrolyzes single- or double-stranded DNA molecules. Digestion yields a mixture of mono- or oligonucleotides displaying 5′ phosphate ends. In the presence of Mg^{2+}, DNase I digests each DNA strand at random positions, whereas in the presence of Mn^{2+}, it cuts each strand at roughly symmetrical sites.

15

Preparation of Rabbit Reticulocyte Extract

Induce a reticulose in an adult rabbit by daily subcutaneous injection of a 2% acetylphenylhydrazine solution in saline (25 mM Tris-HCl, pH 7.0; 130 mM NaCl; 7.5 mM $MgCl_2$; 5 mM KCl). Use 0.4 ml per kg for 4 days.

Exsanguinate the animal by heart puncture after sacrifice on the seventh day. Centrifuge the blood at 8,000g for 10 min at 4 °C and eliminate the leukocytes by adsorption on a cellulose resin. Wash the reticulocytes three times in saline, spinning at 8,000g for 10 min. After the last wash, lyse the reticulocytes by addition of four volumes of ddH_2O containing 25 μM hemine.

Recover the lysate by centrifugation at 27,000g for 15 min and treat with 10 μg/ml of *Micrococcus lysodeikticus* nuclease in the presence of 50 μg/ml of creatine phosphokinase, 2 mM DTT, and 0.75 mM $CaCl_2$ in order to eliminate the endogenous mRNA molecules. After 20-min incubation at -20 °C, add 2 mM EGTA. Distribute in 100-μl aliquots, and store in liquid nitrogen.

16

S1 Nuclease

S1 nuclease is a 32-kDa metalloprotease originally purified from *Aspergillus oryzae* that has an absolute requirement for zinc. It degrades single-stranded DNA in 5′ mono- or oligonucleotides. At low concentration, nuclease S1 is incapable of hydrolyzing double-stranded DNA or RNA, or DNA–RNA hybrids. It is thus widely used to measure the annealing of DNA/DNA or DNA/RNA hybrids since it is still able to digest the protruding single-stranded ends or domains within the hybrid. The enzyme is inhibited by chelating agents such as EDTA, by SDS at a concentration greater than 0.5%, and in phosphate buffers.

17

Reverse Transcriptase

Reverse transcriptase was initially isolated from RNA viruses responsible for specific types of tumors. The most widely used source was the virus for avian myeloblastosis. The enzyme is formed by two subunits α (M_r = 65,000) and β (M_r = 95,000) that associate into an $\alpha\beta$ holoenzyme. The α subunit is responsible for the polymerase activity and requires a RNA template as well as a primer for activity. Reverse transcriptase is mainly used to transcribe mRNA into single-stranded cDNA. Its orthodox polymerase activity is employed to obtain double-stranded from single-stranded cDNA. RNase H is associated to a 24-kDa fragment of the α subunit by enzymatic proteolysis.

18

T4 DNA Ligase

Bacteriophage T4 DNA ligase catalyzes the formation of phosphodiester bonds between the 3'-OH and 5'-P ends of the same DNA strand. It requires Mg^{2+} and ATP as cofactors.

The purification of the enzyme was greatly facilitated by the creation of a recombinant λ bacteriophage containing the T4 DNA ligase gene. This system allowed overproduction of the enzyme when the host was grown at nonpermissive temperature (42 °C).

The ligation reaction occurs in a three-step process. First, the adenylyl group of ATP is transferred in ϵ-NH_2 of the enzyme's lysine residue. In a second step, the group is transferred from the enzyme to the 5'-phosphoryl terminus of the DNA. Lastly, a γ phosphodiester bond is formed between the adenylated 3'-OH and 5'-P extremities and AMP is released.

An enzyme unit is defined as the amount necessary to catalyze the exchange of 1 nmol of NaPPi in ATP at 37 °C for 20 min. For instance, 0.05 units of the enzyme in a 20-µl reaction volume can convert more than 98% of 1 µg of *Hin*dIII-digested λ DNA into ligated forms. Commercially available preparations are devoid of exo- and endonucleases.

19

E. coli Alkaline Phosphatase

E. coli alkaline phosphatase (which has similar properties to calf intestine alkaline phosphatase) is an enzyme able to eliminate the 5′-phosphate groups from substrates such as DNA, RNA, and triphosphate ribonucleosides. The result is a 5′-hydroxyl group that can be labeled with ^{32}P using polynucleotide kinase in the presence of $[\gamma^{32}P]$-ATP. A unit is defined as the amount of enzyme capable of hydrolyzing 1 nmol of $[\gamma^{32}P]$-ATP in 30 min at 37 °C.

Alkaline phosphatase is also used to eliminate the 5′-phosphate groups of linearized vector DNA in order to prevent recircularization or the formation of dimers during ligation reactions. In some instances it is more advantageous to dephosphorylate the passenger DNA rather than the vector DNA in order to avoid inserting more than one extraneous DNA fragment per vector.

Phosphatases are very difficult to eliminate from the reaction medium. Thus they are generally inactivated with EGTA (which chelates Zn^{2+}, a required cofactor) and removed by phenol–chloroform extraction.

20

Deoxynucleotidyl Terminal Transferase (Bollum, 1974)

This enzyme is extracted from the thymus and catalyzes the addition of deoxynucleotides to the 3′-OH terminus of DNA molecules. Terminal transferase is a small ($M_r = 34,000$), basic protein containing two subunits ($M_r = 26,000$ and 8,000).

The enzyme acts as a DNA polymerase by catalyzing the polymerization of 5′-deoxynucleotide triphosphates inthe 5′ → 3′ direction. It is able to act on double-stranded DNA if a 3′ protruding end is available as well as catalyze a limited polymerization of ribonucleosides at the 3′ end of oligodeoxynucleotides. Finally, terminal transferase can use blunt-ended double-stranded DNA as a primer for the addition of ribo- or deoxyribonucleotide phosphate when the cofactor Mg^{2+} is replaced by Co^{2+}.

20

Deoxynucleotidyl Terminal Transferase (Bollum, 1974)

The Polymerase Chain Reaction

The polymerase chain reaction (PCR) is a process designed to amplify a DNA segment located between two regions of known sequence. It involves two 20–24 bp-long oligonucleotides primers that are able to hybridize to DNA sequences lying on opposite strands but flanking the fragment of interest, and a DNA polymerase. In a first step, the DNA is heat-denatured in the presence of an excess of dNTPs and a large molar excess of the two primers. The mixture is then cooled to a temperature which allows the primers to hybridize to their target sequences on the DNA. The primers are then extended—using the single-stranded DNA as a template—by the action of a DNA polymerase. The cycle of heat denaturation, cooling, annealing and extension is repeated 20 to 30 times. Since each newly generated DNA strand can be used as a template by the primers, each successive round doubled the amount of the desired DNA product.

Early protocols made use of the Klenow fragment of *E. coli* DNA polmease I for the extension step. However, since the heating step required for DNA denaturation also inactivated the enzyme, its use has been superseded by that of DNA polymerases isolated from thermophilic microorganisms (e.g., the *Thermus aquaticus* DNA polymerase, Saiki et al., 1988). The *Taq* DNA polymerase is able to survive extended incubation at 95 °C and is active at 72 °C, temperature at which the polymerization step is usually performed. This high temperature guarantees that only legitimate annealing occurs between the primers and the template DNA and thus limits mispriming. This technique is very powerful and it is not rare to obtain μg amounts of DNA starting with picogram quantities. PCR has found applications in the amplification of DNA for sequencing, the generation of probes, mutations, cDNA libraries and new uses are being described monthly.

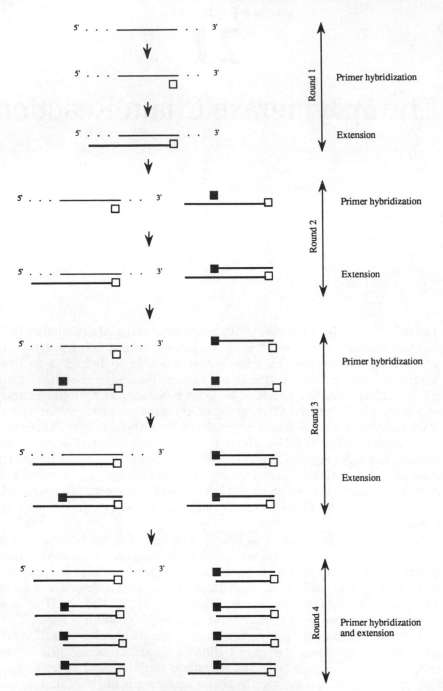

Schematic diagram of the first few rounds of the PCR process. For simplification purposes, only a single-stranded DNA template (top) is shown. The primers are represented by white and black boxes. During the initial rounds of amplification the newly synthesized DNA molecules are heterogenous in size. As more replication rounds take place, the DNA region located between the primers is preferentially amplified and becomes the major product of the reaction.

22

Examples of Adaptor Molecules of Different Sizes

Arrows show the location of the first restriction enzyme site.

Adapter sequence	Restriction sites
5'd (C-G-C-G)3'	*Tha*I
5'd (G-C-G-G-C)3'	Hh*A*I
5'd (C-C-C-G-G-G)3'	*Hpa*II , *Sma*I
5'd (G-G-A-A-T-T-C-C)3'	*Eco*RI
5'd (C-A-A-G-C-T-T-G-)3'	*Hind*III
5'd (C-G-G-A-T-C-C-G)3'	*Bam*HI
5'd (C-C-G-A-A-T-T-C-G-G)3'	*Eco*RI
5'd (C-C-A-A-G-C-T-T-G-G)3'	*Hind*III
5'd (C-C-G-G-A-T-C-C-G-G)3'	*Bam*HI
5'd (C-C-C-G-A-A-T-T-C-G-G-G)3'	*Eco*RI
5'd (C-C-C-A-A-G-C-T-T-G-G-G)3'	*Hind*III
5'd (C-C-C-G-G-A-T-C-C-G-G-G)3'	*Bam*HI

23

Ribonuclease H

This enzyme can specifically degrade RNA when it is hybridized to DNA. One unit of ribonuclease H is defined as the amount required to hydrolyze 1 nmol of [^{3}H]-poly(A) in [^{3}H]-poly(A):poly(dT) in 20 min at 37 °C.

24

Characteristics of a Few Commonly Used Cosmids

Appendix 24 Characteristics Of A Few Commonly Used Cosmids

Name	Replicon	Size (kbp)	Resistances	Cloning Sites	Cloning Capacity (kbp)
pJC74	ColE1	15.8	Ampicillin	*Eco*RI *Bam*HI *Bgl*II *Sal*I	21–37
pJC81	pMB1	7.1	Ampicillin Tetracycline	*Kpn*I *Bam*HI *Hind*III *Sal*I	30–46
pHC79	pMB1	6.4	Ampicillin Tetracycline	*Eco*RI *Cla*I *Hind*III *Bam*HI *Sal*I *Pst*I	31–47
pMu-A-10	pMB1	4.8	Tetracycline	*Eco*RI *Bal*I *Pvu*I *Pvu*II	32–48
pJB8	ColE1	5.4	Ampicillin	*Bam*HI *Hind*III *Sal*I	31–47
pH262	ColE1	2.8	Kanamycin	*Bam*HI *Eco*RI *Hinc*II	34–50
pMF7	pMB1	5.4	Ampicillin	*Eco*RI *Sal*I	32–48

Answers

Chapter 1

1.1: Strain 1: Ap^s, Tc^s, Cm^s; strain 2: Ap^r, Tc^r, Cm^r; strains 3 and 4: Ap^r, Tc^s, Cm^s.

1.2: Bacterial growth is measured by recording the spectrophotometric absorbance of a bacterial cell culture at 600 nm. On a semilogarithmic plot of OD_{600} vs. time, the exponential phase corresponds to a straight line. The doubling time is determined on this line and corresponds to the time necessary to double the OD_{600} value.

Chapter 2

2.1: *Hha*I yields a 3′ overhang while the three other restriction enzymes give protruding 5′ ends. (*Hpa*II gives a 2-, *Asu*I a 3-, and *Eco*RII a 5-nucleotide extension.)

2.2: *Eco*RII recognizes two sequences (CCAGG and CCTGG), *Scr*FI four (CCAGG, CCTGG, CCCGG, and CCGGG), and *Afl*III four (ACACGT, ACATGT, ACGCGT, and ACGTGT).

2.3: All three enzymes follow the symmetry rule but contain an internal (*Eco*RII and *Hinc*II) or external (*Hae*II) degenerescence.

2.4: *Xho*I is an ioschizomer of *Blu*I, and *Sma*I an isoschizomer of *Xma*I. *Xho*I and *Blu*I have the same recognition sequence and cleavage sites: they are true isoschizomers. The two other enzymes have the same recognition sites but different cleavage sites and are imperfect isoschizomers.

2.5: Tetrameric sequences theoretically occur $1/4^4$—that is, once every 256 bp. Hexameric recognition sequences occur $1/4^6$—that is, once every 4,096 bp.

2.6: Sixteen for tetrameric cutters and 64 for hexameric cutters. (See Table 2.3.)

2.7: Restriction enzymes that recognize $G-C$ pairs are found more frequently than those recognizing $A-T$ pairs. This is explained by the fact that $G-C$ bonds contain three hydrogen bonds while $A-T$ bonds contain two hydrogen bonds. Hence, the $G-C$-rich sequences are more stable than the $A-T$-rich sequences.

2.8:

$$a = \frac{\text{Molecular weight of } \lambda}{\text{No. restriction sites in } \lambda} \times \frac{\text{No. restriction sites in } P}{\text{Molecular weight of } P}$$

Chapter 3

3.1: 5'-TCGA-3' and 5'-GGATCC-3'.

3.2: *Sau*3A cuts GATC and G$\overline{A}$TC; *Mbo*I only cuts GATC, and *Dpn*I only cuts GATC.

3.3: ATCGATC for *Cla*I; TCTAGATC for *Xba*I; GGTGATC for *Hph*I.

3.4: The *Msp*I methylase methylates the first cytosine of the CCGG sequence preceding the first *Bam*HI site and prevents the enzyme from cutting at this position. *Bam*HI will, however, still be able to cut in its second site at the 3' end of the sequence.

3.5: The *Sma*I recognition sequence contains the CG doublet, which frequently displays a methylated cytosine that prevents digestion. In addition this pair of nucleotides, and thus the *Sma*I recognition sequence, does not occur frequently on the genome (probably as a result of hypermutability of the methylated cytosine).

Chapter 4

4.1: The top curve of *Answers* Fig. 1 shows that the electrophoretic mobility of DNA fragments increases as their size decreases. When a semilog plot is used (bottom) there is a rough inverse proportionality between these values in the 20–3-kbp range. A similar relationship, but with different slopes, is observed for the smaller DNA fragments. It is possible to use such correlations to determine the size of the DNA fragments in a sample.

4.2: In the case of acrylamide gels and with reference to the question of whether DNA fragments or proteins are resolved, there is a strict inverse proportionality between the two values. As a consequence, an accurate determination of molecular weights is possible. The amount of ethidium bromide inserted between the DNA bases is proportional to the size of a fragment. For the particular band marked by an asterisk in Fig. 4.11, the intensity should be much

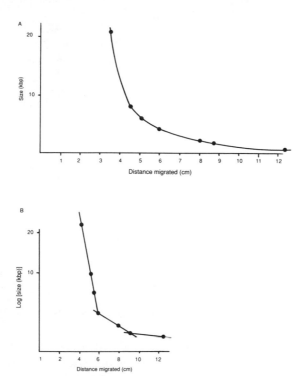

Figure 1. Graphic solution for problem 4.1.

lower; this indicates that this band corresponds to two DNA fragments very close in molecular weight.

4.3: *Eco*RI cuts on the average once every 4,000 bp on the genome. This yields 10^5–10^6 fragments per genome. When the digested genomic DNA is run on an agarose gel a continuous streak corresponding to a multitude of fragments of different molecular weights is observed. The digestion is considered to be complete (all sites hydrolyzed) when the streak is uniform in intensity. The discrete bands labeled S on Fig. 4.9 correspond to highly repetitive DNA sequences containing at least one *Eco*RI site.

Chapter 5

5.1: Upon infection bacteriophage λ injects its DNA in the cytoplasm of the cell (Fig. 2) and θ replication starts. Concatemers can be formed by rolling-circle replication or multimers generated by recombination. These two types of molecules are the only ones that can be encapsidated.

5.2: Preliminary incubation at room temperature favors phase adsorption while a shift to 37 °C initially helps the injection of phage DNA in the cell. The

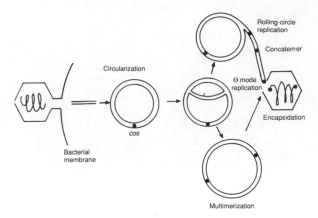

Figure 2.   The lytic cycle of bacteriophage λ (problem 5.1)

preliminary incubation synchronizes infection and reduces the heterogeneity in plaque size.

5.3: As a general rule, any factor that increases the time for the generation of a confluent bacterial layer will also increase plaque size. Thus large plaques are obtained if few cells are mixed with the top agar and the phage particles; if the nutrient medium used is less rich; if the plates have been freshly prepared; if growth is carried out under high humidity conditions; or if the agar concentration is low.

Chapter 6

6.1: (1) For *Eco*RI, *Hin*dIII, or *Sal*I insertion: 0–9.1 kbp may be cloned. (2) For *Eco*RI-*Hin*dIII substitution, from 0 to 11 kbp; for *Hin*dIII-*Sal*I substitution, from 6.1 to 19.1 kpb; for *Eco*RI-*Sal*I substitution, from 8.1 to 21 kbp.

6.2: (1) The parental vector will express β-galactosidase and generate blue plaques in the presence of X-gal. The recombinants will contain an inactive β-galactosidase as a result of the insertional inactivation by the passenger DNA of the *lac*Z gene and the plaques will be colorless. (2) The vector will form blue plaques through titration of the *lac*I repressor; the recombinant will yield colorless plaques since the *lac* operator is inactivated. (3) The parental will give blue plaques since the amber mutation in *lac*Z will be suppressed by the vector-encoded *sup* genes; the recombinant will give colorless plaques.

6.3: The *Bam*HI restriction site is G↓GATCC and allows cloning of DNA fragments containing a 5′-GATC overhang. These can be generated by a number of enzymes [e.g., *Bgl*II (A↓GATCT), *Bcl*I (T↓GATCA), and *Xho*I (Pu↓GATCPy)].

Chapter 7

7.1: Plasmid useful as cloning vectors must:

- Autonomously replicate for easy isolation and amplification
- Encode gene(s) for antibiotic resistance that confer an easily selectable phenotype
- Contain a restriction site in a nonessential region of the genome and preferentially in an antibiotic resistance gene where the passenger DNA can be cloned

7.2: Small plasmids are abundant within the cell and easier to manipulate. Their restriction map is not as complicated relative to larger plasmids that also encode a larger number of restriction sites.

7.3: Plasmid pBR322 is digested with *Hin*dIII and *Bam*HI under suitable buffer conditions for both enzymes. The large fragment is purified by electrophoresis on low-melting-point agarose and ligated with a passenger DNA fragment displaying compatible overhangs. Since the extremities generated by *Bam*HI and *Hin*dIII digestion are not compatible, the vector DNA will be unable to religate to itself, which reduces background.

7.4:

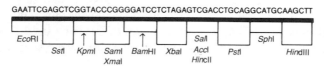

GAATTCGAGCTCGGTACCCGGGGATCCTCTAGAGTCGACCTGCAGGCATGCAAGCTT

EcoRI — Sstl — Kpml — Saml — BamHI — Xbal — Accl / Hincll — Sal — Pstl — Sphl — HindIII

Xmal

Figure 3. Solution to problem 7.4.

Chapter 8

8.1: Bacterial DNA may interfere with plasmid DNA during electrophoresis, transformation, and hybridization experiments. The degree of contamination may be estimated by electrophoresis. The transformation yield is higher if purified plasmid DNA is used. Genomic DNA can interfere with the hybridization of probes to recombinant plasmid DNA.

8.2: Plasmid DNA can adopt three main conformations (*Answer* Fig. 4): linearized (e.g., following restriction-enzyme digestion), nicked (when one of the strands is broken), or supercoiled (where the DNA contains multiple twists). The different conformations influence electrophoresis and sedimentation patterns, the ligation to other DNA fragments, and the susceptibility to denaturation.

8.3: The clarification of the cell lysate by centrifugation eliminates most of the genomic DNA from the preparation. In addition, the plasmid DNA is usually amplified and is present in large amounts. The linear bacterial DNA incorpo-

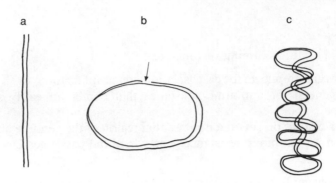

Figure 4. The three different conformations of plasmid DNA: (a) linearized, (b) nicked, (c) supercoiled.

rates much more ethidium bromide than the supercoiled plasmid DNA, which gives the latter a higher apparent density in the gradient.

8.4: *Answer* Fig. 5 shows that for the same radial distances r_a and r_b, the actual distances between the bands (d_1 and d_2) are greater in the case of a fixed-angle rotor.

8.5: Using electrophoresis on an agarose gel it is possible to determine whether the plasmid DNA is contaminated by high-molecular-weight genomic DNA or low-molecular-weight RNA. The same technique can be used to investigate plasmid conformation and the efficiency of digestion by a restriction enzyme.

Chapter 9

9.1: *Answer* Fig. 6 shows the steps involved in the transformation of gram-positive cells by extraneous DNA: (1) the cell becomes competent; (2) the double-stranded foreign DNA adsorbs at the surface of the cell; (3) the single-stranded DNA penetrates within the cell; (4) the foreign DNA hybridizes to a homologous section of the host chromosome; (5) the DNA is integrated in the host chromosome by recombination.

9.2: The frequency of transformation is determined as the number of transformants (i.e., Ap^r colonies) among all the viable cells (i.e., that form a colony on a LB plate in the absence of selective pressure) per μg of DNA used in the transformation. The presence of the plasmid in cells growing on ampicillin plates is verified by miniprep and agarose electrophoresis.

9.3: 10^6 per μg.

9.4: Transform competent cells with the plasmid of choice. Do serial dilutions of the transformed cells in a small volume of LB medium. Plate on petri dishes containing nutrient agar in the presence or absence of selective pressure. Count the colonies after overnight incubation.

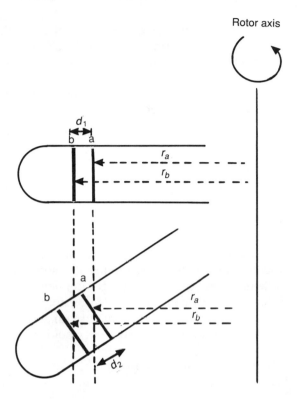

Figure 5. This schematic representation shows that $d_2 > d_1$ in a fixed-angle rotor for identical radial distances r_a and r_b.

Chapter 10

10.1: Streak the strain on two LB plates. Incubate one at 30 °C and the other at 42 °C overnight. Colonies should be obvious on the former since there is no induction. In contrast, no colonies should be present on the 42 °C plate since incubation at this temperature induces prophages.

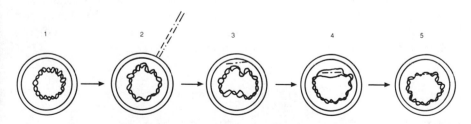

Figure 6. Different steps in the transformation process.

10.2: Table 10.2 shows that an increase in the DNA concentration also enhances the number of plaques formed but that the efficiency of encapsidation remains constant (about 10^7) even when the amount of substrate is multiplied by 40.

10.3: In this case there is a selectivity according to the size of the DNA. Large genomes are encapsidated with a higher efficiency.

Chapter 11

11.1: The nucleic acid concentration is estimated by the absorbance at 260 nm. A ratio of $A_{260}:A_{280}$ close to 2 indicates that the proteins have been properly removed from the sample. An A_{320} 10% lower than A_{260} indicates that the DNA has been well purified and that there are no aggregates. On the spectrum the A_{230} value cannot be used to determine the absence of histones and other proteins around 230–240 nm since the dialysis buffer also absorbs in this region.

11.2: The large difference between the concentration determined by spectrophotometry and the true DNA concentration measured by chemical dosage indicates the presence of a large amount of contaminating RNA in the sample. Specific dosage of RNA concentration with ocrinol would give the extent of the contamination.

11.3 Count the cells after isolation and washing of the leukocytes. The amount of DNA per cell is about 6 pg.

Chapter 13

13.1: The absorbance pattern at 260 nm shows that the RNA was essentially not degraded since there is about twice as much 28S as 18S RNA. The distribution of the mRNA is heterogeneous and extends to the 25 to 28S region of the gradient. The small amount of polyadenylated DNA in the 4 to 7S region of the gradient indicates that the preparation is of good quality.

13.2: The concentration is 1.27 mg/ml, and the quantity is 25.4 mg.

13.3: 0.6% of the total RNA is poly(A); the poly(A) only represents 0.047% of the RNA in the column flow-through but 7.73% in the bound fraction. This single chromatography step corresponds to a purification factor of 12.88.

13.4: The poly(A)$^+$ mRNA is contaminated by ribosomal RNA. This is not totally unexpected since these RNA species can form aggregates with poly(A) mRNA and thus bind to an oligo(dT) column.

Chapter 15

15.1: For both extracts, K^+ can significantly influence the translational activity of the system and totally inactivate the extract at high concentration.

15.2: Methionine is commercially available at high specific activity. Proteins labeled with ^{35}S give a stronger and faster signal on autoradiograms compared to ^{14}C- or ^{3}H-labeled proteins. Table 15.1 further shows that there is relatively little methionine in unfractionated rabbit reticulocyte extract, which prevents dilution of the label and results in better incorporation.

15.3: The endogenous translational activity is measured by running a parallel experiment in the absence of exogenous mRNA. The amount of protein synthesized is determined by TCA precipitating an aliquot and counting. The incorporation of [^{35}S]-methionine is reported in cpm/μl of sample. The amount of protein immunoprecipitated with specific antibodies is also done by scintillation counting.

15.4: 11.6%. No; it is usually necessary to eliminate some of the products that could co-precipitate with the specific protein recognized by the antibody. Specific clarifying systems can remove 5–10% of the incorporated radioactivity.

15.5: The main band in the gel migrates at the expected position of albumin. The low-molecular-weight products probably correspond to partially synthesized albumin molecules that have been co-precipitates by the antibodies. The reason the newly synthesized albumin is heavier than plasmatic albumin is that the former is still in the prepro form and includes the sequences that are normally removed in the maturation process.

Chapter 16

16.1: The polysome preparation appears to be intact with most polysomes sedimenting at a position corresponding to at least 10 monosomes.

16.2: Even under denaturing conditions, there is translational activity in the 28S region of the gradient; this is indicative of the formation of aggregates.

16.3: Fivefold.

Chapter 17

17.1: The cDNA peak corresponds to 8,962 cpm for a total radioactivity in all fractions of 736,982 cpm. Thus 1.216% of the label has been incorporated into growing cDNA strands. There is 46.7 μg of dCTP in the 1-ml sample and 0.568 μg has been incorporated. Assuming that C appears once every four nucleotides, 2.271 μg of cDNA has been labeled. This corresponds to a yield of mRNA, relative to single-stranded cDNA, of 22.7%.

17.2: From 2 to 5 μg of poly(A)$^+$ mRNA.

17.3: After treatment with S1 nuclease, the average size of the double-stranded cDNA population decreases.

17.4: The size distribution of ^{32}P-labeled cDNA varies between 500 and 1,800 nucleotides, which is in good agreement with the length of polyadenylated messengers prepared from liver. The discrete bands correspond to either messages produced in large amount or to incomplete transcripts at given points along the mRNA molecules.

17.5: There are three main reasons for obtaining incomplete double-stranded cDNA. First, synthesis of the first strand is incomplete. (This gives a missing 5' region.) Second, the first strand has been copied imperfectly (missing 3' region). Third, only the central region of the mRNA is obtained if the copy of an incomplete first strand is defective.

Chapter 18

18.1: *Answer* Fig. 7 shows a schematic representation of the ligations between the 3'-OH and 5'-P DNA extremities that are located on one strand only (nick), two strands at different positions (staggered), or two strands and facing each other (blunt).

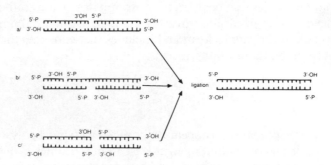

Figure 7.   Ligation of: a) a nick in one of the DNA strands; b) a staggered-ended cut; c) a blunt-ended cut.

18.2: The 4-bp overhang generated by *Pst*I digestion (CTGCA↓G) will generate only two weak A−T bonds upon annealing to the complementary strand, while that obtained with *Eco*RI (G↓AATTC) will yield four less-stable A−T bonds.

18.3: The same TCGA extension is generated following digestion.

18.4: For pBR322: $(2 \times 1 \times 10^6)/(2.9 \times 10^6) = 0.69$ pmol of ends per μg of plasmid. For λ:

$(2 \times [18 + 1] \times 10^6)/(3.2 \times 10^7) = 1.19$ pmol of ends per μg of phage.

18.5: The effective concentration can be obtained using $j = \left(\dfrac{3}{2 \cdot \pi \cdot l \cdot b} \right)^{3/2}$

The result is in ends per ml. In the formula, l is the length of the DNA molecule, and b is the minimal length of the DNA able to circularize. In the case of bacteriophage λ, $l = 13.2$ μm, $b = 7.7 \times 10^{-2}$ μm; thus $j = 3.6 \times 10^{11}$ ends per ml.

18.6: At 10 μg/ml, $j/i = 5$ and the formation of circular monomers will be favored. At 50 μg/ml, $j/i = 0.5$ and linear concatemers will be formed.

Chapter 19

19.1:
Number of residues added =

$$\frac{\text{cpm per } \mu\text{l of acid-insoluble material}}{\text{cpm per molecule of dNTP} \times \text{total number of 3' ends per } \mu\text{l of reaction mix}}$$

The numerator value is obtained by measuring the cpm precipitated with cold TCA. The number of cpm per dNTP is obtained by:

$$\frac{\text{cpm per } \mu\text{l of [}^{32}\text{P]-dNTP}}{\text{Concentration in mol/ml} \times \text{Avogadro number of dNTPs}}$$

The total number of 3' end per μl of mix is given by:

$$\frac{[\text{Amount of DNA added (in } \mu\text{g)]} \cdot (6.02 \cdot 10^{23}) \cdot 2}{[\text{Molecular weight of DNA (in g)]} \cdot 10^6 \cdot (\text{reaction volume})}$$

19.2: Figure 8

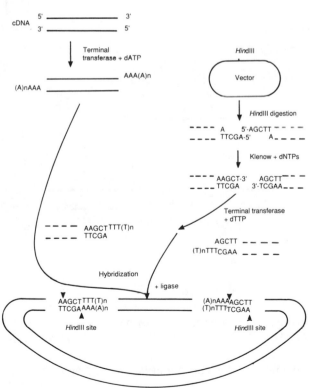

19.3: One picomole of plasmid represents 2.5 μg and, consequently, 1 pmol of 3′ *Pst*I ends will correspond to 1.25 μg. Under the conditions described 5 μCi corresponds to 5 pmol. Addition of 5 μCi of [^{32}P]-dCTP to 1 μl of 0.5 mM dCTP will give 505 pmol. For [^{3}H]-dCTP, 5 μCi corresponds to 295 pmol. Addition of 5 μCi to 0.5 mM of dCTP will yield 495 pmol.

Chapter 20

20.1: Figure 9

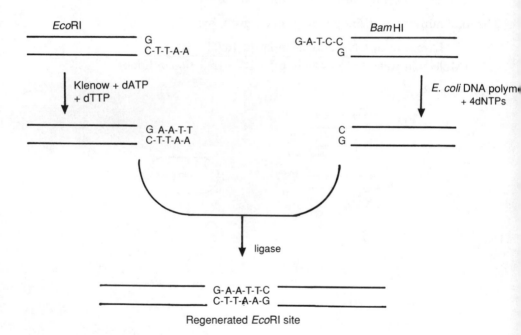

Regenerated *Eco*RI site

20.2: The linker is labeled with [α^{32}P]-ATP at its 5′ end using polynucleotide kinase. The products of the reaction are resolved by agarose gel electrophoresis under denaturing conditions and visualized by autoradiography.

20.3: CATCGATG: *Cla*I, *Taq*I, TflI; GTGGCCAC: *Bal*I, *Bsu*RI, *Cfr*I, *Hae*I, *Hae*II.

20.4: *Alu*I. Hexameric linkers do not alter the reading frame.

20.5: Figure 10

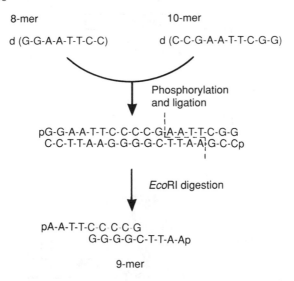

20.6: Figure 11

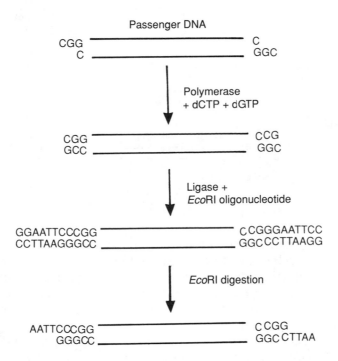

20.7: *Eco*RI and *Hha*I for the first adaptor; *Bam*HI and *Sma*I for the second.

20.8: Figure 12

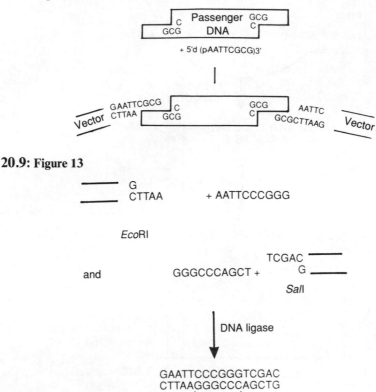

20.9: Figure 13

20.10: Figure 14

The new site generated is *Xho*I

Chapter 21

21.1: An ideal genomic library should contain all the DNA sequences in the genomic DNA. Each cloned fragment should be large enough to contain an entire gene and its adjacent sequences—but not too large, so that it can still be mapped with restriction enzymes. The library should contain a reasonable number of clones for easy manipulation that display overlapping regions to allow chromosome walking. It should also be stable upon storage.

21.2: Since an *Eco*RI fragment has an average size of 4 kpb, about 7×10^5 independent recombinants must be prepared.

21.3: 1.4×10^5.

21.4: For $P = 0.95$,

$$N = \frac{\ln(1 - 0.95)}{\ln\left(1 - \dfrac{1}{1.4 \times 10^5}\right)} = 4.2 \times 10^5$$

For $P = 0.99$, $N = 6.5 \times 10^5$, a much higher number.

21.5: *E. coli*, 9.2×10^2; *S. cerevisiae*, 3.2×10^3; *Drosophila*, 3.2×10^4.

21.6: Assuming a maximum efficiency and no loss during any stage of the cloning, 1 μg of DNA fractionated to the appropriate size will yield a library of at least 10^6 recombinants. Such a high number is, however, rarely reached in practice.

21.7: These vectors have been designed to reduce the background of nonrecombinant phages. The multiple cloning sites are located between the arms of the phage and the central fragment, and it is possible to prevent the stuffer from religating to the arms using double digestion. The Spi$^+$/P2 selection to discriminate against non recombinants is also possible.

Chapter 22

22.1: The cDNA libraries are smaller than genomic libraries (i.e., they consist of a smaller number of clones). Consequently they are easier to manipulate. The difference in library size results from the fact that a large proportion of eukaryotic DNA does not encode genes. In addition, only the exons are translated. The cDNA used in library construction does not contain introns and adjacent sequences. It is also possible to prepare cDNA libraries from specific mRNAs or tissues, or at different development stages. Finally, it is only possible to obtain the genome of certain RNA viruses through conversion into cDNA and construction of a library.

22.2: It is very difficult to obtain full-length cDNA molecules since the mRNA contains secondary structures that prevent the progression of reverse transcriptase. This efficiency of the reaction can be improved by increasing the incubation temperature, by denaturing the mRNA using heat or methylmercury hydroxide treatment, or by using random primers.

22.3: The pUC vectors contain a multiple cloning site; they can be amplified and are smaller than the pBR-based vectors. In addition, cloning can be performed in the β-galactosidase gene, which permits a direct color selection of the recombinants.

22.4: The length of the second cDNA strand is identical to that of the template cDNA strand. In the classic synthesis reaction the double-stranded denatured cDNA is twice as long.

22.5: λgt10 is a vigorous phage that gives large plaques uniform in size and appearance. Recombinant phages can be selected using radiolabeled probes. λgt11 is an expression vector.

Chapter 23

23.1: In a 25-kbp cosmid, the capacity ranges between 11 and 27 kbp; in an 11-kbp cosmid, from 16 to 42 kbp; in a 3-kbp cosmid, from 34 to 50 kpb.

23.2: The restriction site of *Sau*3A ($\downarrow$GATC) contains one of each possible nucleotide; thus the enzyme cuts randomly along genomic DNA. In addition, the enzyme recognition sequence does not include a GC pair (which is rare in eukaryotic DNA), and the distribution of the sites is therefore not affected. Finally, the overhang it generates is compatible with *Bam*HI present on a number of cosmids.

23.3: 3.22×10^5 recombinants for a cosmid library complete at 99% but 8.59×10^5 for a phage library complete at 99%.

References

Chapter 1

Bachmann, B. J. (1972) *Bact. Rev.* **36:** 525
Demerec, M., Adelberg, E. A., Clark, A. J. and Hartman, P. E. (1966) *Genetics,* **54:** 61

Chapter 2

Dussoix, D. and Arber, W. (1962) *J. Mol. Biol.,* **5:** 37
Kelly, T. J. and Smith, H. O. (1970) *J. Mol. Biol.,* **51:** 393
Meselson, M. and Yuan, R. (1968) *Nature,* **217:** 1110
Smith, H. O. and Wilcox, K. W. (1970) *J. Mol. Biol.,* **51:** 379

Chapter 3

Greene, P. J., Poonian M. S., Nussbaum, A. L., Tobias, L., Garfin, D. E., Boyer, H. W. and Goodman, H. M. (1975) *J. Mol. Biol.,* **99:** 237
Marinus, M. G. and Morris, N. R. (1973) *J. Bacteriol.,* **114:** 1143
May, M. S. and Hattman, S. (1975) *J. Bacteriol.,* **123:** 768
Waalwijk, C. and Flavell, R. A. (1978) *Nucleic Acids Res.,* **5:** 3231

Chapter 4

Aaij, C. and Borst, P. (1972) *Biochem. Biophys. Acta,* **269:** 192
Brunk, C. F. and Simpson, L. (1977) *Anal. Biochem,* **82:** 455

Ghrayeb, J., Kimura, H., Takahara, M., Hsiung, H., Masui, Y. and Inouye, M. (1984) *EMBO J.,* **3:** 2437

Kopchick, B. R., Cullen, B. R. and Stacey, C. W. (1981) *Anal. Biochem.,* **115:** 419

Loening, V. E. (1967) *Biochem. J.,* **102:** 251

Maniatis, T., Fritsch, E. F., and Sambrook, J. (1982). Molecular cloning: a laboratory manual. Cold Spring Harbor Laboratory Press, Cold Spring Harbor, N.Y.

Sharp, R. A., Sugden, B. and Sambrook, J. (1973) *Biochemistry,* **12:** 3055

Chapter 5

Sanger, F., Coulson, A. R., Hong, G. F., Hill, D. F. and Petersen, G. B. (1982) *J. Mol. Biol.,* **162:** 729

Chapter 6

Blattner, F. R., Williams, B. G., Blechl, A. E., Deniston-Thompson, K., Faber, H. E., Furlong, L. A., Grunwall, D. J., Kiefer, D. O., Moore, D. D., Shumm, J. W., Sheldom, E. L. and Smithies, O. (1977) *Science,* **196:** 164

Davison, J., Brunel, F. and Merchez, M. (1979) *Gene,* **8:** 69

Karn, J., Brenner, S., Barnett, L. and Cesareni, G. (1980) *PNAS,* **77:** 5172

Leder, P., Tiemeir, D. and Enquist, L. (1977) *Science,* **196:** 175

Loenen, W.A.M. and Brammar, W. J. (1980) *Gene,* **10:** 249

Loenen, W.A.M. and Blattner, F. R. (1983) *Gene,* **26:** 171

Maniatis, T., Fritsch, E. F., and Sambrook, J. (1982). Molecular cloning: a laboratory manual. Cold Spring Harbor Laboratory Press, Cold Spring Harbor, N.Y.

Murray, N. E. and Murray, K. (1974) *Nature,* **251:** 476

Murray, N. E., Brammar, W. J. and Murray, K. (1977) *Mol. Gen. Genet.,* **150:** 53

Nathans, J. and Hogness, D. S. (1983) *Cell,* **34:** 807

Rambach, A. and Tiollais, P. (1974) *PNAS,* **71:** 3927

Rimm, D., Horness, D., Kucera, J. and Blattner, F. R. (1980) *Gene,* **12:** 301

Szybalski, E. H. and Szybalski, W. (1979) *Gene,* **7:** 217

Thomas, M., Cameron, J. R. and Davis, R. W. (1974) *PNAS,* **71:** 4579

Weil, J., Cunningham, R., Martin, R., Mitchell, E. and Bolling, B. (1973) *Virology,* **50:** 373

Yamamoto, K. R., Alberts, B. M., Benzinger, R., Lawhorne, L. and Treiber, G. (1970) *Virology,* **40:** 734

Chapter 7

Backman, K. and Boyer, H. W. (1983) *Gene,* **26:** 197

Bochner, B. R., Huang, H. C., Schieven, G. L. and Ames, B. N. (1980) *J. Bacteriol.,* **143:** 926

Bolivar, F., Rodriguez, R. L., Betlach, M. V. and Boyer, H. W. (1977a) *Gene,* **2:** 75

Bolivar, F., Rodriguez, R. L., Greene, P. J., Betlach, M. V., Heyneker, H. L., Boyer, H. W., Crosa, J. H. and Falkow, S. (1977b) *Gene,* **2:** 95

Clewell, D. B. and Helsinki, D. R. (1972) *J. Bacteriol.,* **110:** 1135

Cohen, S. N. and Chang, A.C.Y. (1977) *J. Bacteriol,* **132:** 734

Dente, L., Cesareni, G. and Cortese, R. (1983) *Nucleic Acids Res.,* **11:** 1929

Dotto, G. P. and Horinchi, K. (1981) *J. Mol. Biol.,* **143:** 169

Hershfield, V., Boyer, H. W., Yanofsky, C., Lovett, M. A. and Helsinki, D. R. (1974) *PNAS,* **71:** 3255

Morrow, J. F., Cohen, S. N., Chang, A.C.Y., Boyer, H. W., Goodman, H. and Helling, R. B. (1974)
    PNAS, **71:** 1743
Norrander, J., Kempe, T. and Messing, J. (1983) *Gene,* **26:** 101
Peden, K.W.C. (1983) *Gene,* **22:** 277
Roberts, T. M., Swanberg, S. L., Poteete A., Riedel G. and Backman K. (1980) *Gene,* **12:** 123
Rüther, V. (1980) *Mol. Gen. Genet.,* **178:** 475
Shumann, W. (1979) *Mol. Gen. Genet.,* **174:** 221
Sutcliffe, J. G. (1978) *Nucleic Acids Res.,* **5:** 2721
Vieira, J. and Messing, J. (1982) *Gene,* **19:** 259
Yanisch-Perron, C., Vieira, J. and Messing, J. (1985) *Gene,* **33:** 103

Chapter 8

Birnboim, H. C. and Doly, J. (1979) *Nucleic Acids Res.,* **7:** 1513
Clewell, D. B. (1972) *J. Bacteriol.,* **110:** 667
Clewell, D. B. and Helinski, D. R. (1969) *PNAS,* **62:** 1159
Holmes, D. S. and Quigley, M. (1981) *Anal. Biochem.,* **114:** 193
Klein, R. D., Selsing, E. and Wells, R. (1980) *Plasmid,* **3:** 38

Chapter 9

Cohen, S. N., Chang, A.C.Y. and Hsu, L. (1972) *PNAS,* **69:** 2110
Dagert, M. and Ehrlich, S. D. (1979) *Gene,* **6:** 23
Henner, W. D., Kleber, I. and Benzinger, R. (1973) *J. Virol.,* **12:** 741
Kushner, S. R. (1978) *in Genetic Engineering,* p. 17
Lederberg, E. M. and Cohen, S. N. (1974) *J. Bacteriol.,* **119:** 1072
Mandel, M. and Higa, A. (1970) *J. Mol. Biol.,* **53:** 159
Morrison, D. A. (1979) *Methods Enzymol.,* **68:** 326
Norgard, M. V., Keem, K. and Monahan, J. J. (1978) *Gene,* **3:** 279

Chapter 10

Becker, A. and Gold, M. (1975) *PNAS,* **73:** 4114
Hohn, B. and Murray, K. (1977) *PNAS,* **74:** 3259
Hohn, B. (1979) *Methods in Enzymology,* **68:** 779
Sternberg, N., Tiemeir, D. and Enquist, L. (1977) *Gene,* **1:** 255

Chapter 11

Blin, N. and Stafford, D. W. (1976) *Nucleic Acids Res.,* **9:** 2303
Gautreau, Ch., Rahuel, C., Cartron, J. P. and Lucotte, G. (1983) *Anal. Biochem.,* **134:** 320

Chapter 12

Banner, D. B. (1982) *Anal. Biochem.,* **125:** 139
Blin, N., Gabain, A. V. and Bujard, H. (1975) *FEBS lett.,* **53:** 84

Chen, C. W. and Thomas, C. A. (1980) *Anal. Biochem.,* **101:** 339
Maxam, A. M. and Gilbert, W. (1977) *PNAS,* **74:** 560
Mc Donnel, M. W., Simon, M. N. and Studier, F. W. (1977) *J. Mol. Biol.,* **110:** 119
Smith, H. O. (1980) *Methods Enzymol,* **65:** 371
Weislander, L. (1979) *Anal. Biochem.,* **98:** 305
Yang, R.C.A., Lis, J. and Wu, R. (1979) *Methods Enzymol,* **68:** 176

Chapter 13

Aviv, H. and Leder, P. (1972) *PNAS,* **69:** 1408
Bantle, J. A., Maxwell, I. H. and Hahn, W. E. (1976) *Anal. Biochem.,* **72:** 413
Bishop, J. O., Rosbach, M. and Evans, D. (1974) *J. Mol. Biol.,* **85:** 75
Chirggwin, J. M., Przybyla, A. E., McDonald, R. J., and Rutter, W. J. (1979) *Biochemistry,* **18:** 5294
Kirby, K. S. (1968) *Methods Enzymol,* **12B:** 87
Lindberg, V. and Persson, T. (1982) *Eur. J. Biochem.,* **31:** 346

Chapter 14

Bailey, J. M. and Davidson, N. (1976) *Anal. Biochem.,* **70:** 75
Lehrach, H., Diamond, D., Wozney, J. M. and Boedtker, H. (1977) *Biochemistry,* **16:** 4743
Lemischka, I. H., Farmer, S., Racaniello, V. R. and
    McMaster, G. K. and Carmichael, G. G. (1977) *PNAS,* **74:** 4835
Rave, N., Crkvenjakov, R. and Boedtker, H. (1979) *Nucleic Acids Res.,* **6:** 3559

Chapter 15

Pelham, H.R.B. and Jackson, R. J. (1976) *Eur. J. Biochem.,* **67:** 247
Roberts, B. E. and Paterson, B. B. (1973) *PNAS,* **70:** 2330

Chapter 16

Schimke, R. T., Palacios, R., Sullivan, O., Kicly, M. C., Gonzales, C. and Taylor, J. A. (1974) *Methods Enzymol,* **30:** 631

Chapter 17

Efstratiadis, A., Kafatos, F. C., Maxam, A. M. and Maniatis, T. (1976) *Cell,* **7:** 279
Gething, M. J., Bye, J., Skehel, J. and Waterfield, M. (1980) *Nature,* **287:** 301
Humphries, P., Cochet, M., Krust, A., Gerlinger, P., Kourilsky, Ph. and Chambon, P. (1977) *NAR,* **4:** 2389
Retzel, E. F., Collet, M. S. and Faran, A. T. (1980) *Biochemistry,* **19:** 513
Rougeon, F. and Mach, B. (1976) *PNAS,* **73:** 3418
Ullrich, A., Shine, J., Chirgwin, J., Pictet, R., Fisher, E., Rutter, W. J. and Goodman, H. M. (1977) *Science,* **196:** 1313
Wickens, M. P., Buell, G. N. and Schimke, R. T. (1978) *J. Biol. Chem.,* **253:** 2483

Chapter 18

Dugaiczyk, A., Boyer, H. W. and Goodman, A. M. (1975) *J. Mol. Biol.,* **96:** 171
Ullrich, A., Shine, J., Chirgwin, J., Pictet, R., Tisher, G., Rutter, W. J. and Goodman, H. (1977) *Science,* **196:** 1313

Chapter 19

Brutlag, D., Fry, K., Nelson, T. and Hung, P. (1977) *Cell,* **10:** 509
Jackson, D. A., Symons, R. H. and Berg, P. (1972) *PNAS,* **69:** 2904
Land, H., Grey, M., Hanser, H., Lindenmaier, W. and Shutz, G. (1981) *Nucleic Acids Res.,* **9:** 2251
Peacock, S. L., Mc Iver, C. M. and Monohan, J. J. (1981) *Biochim. Biophys. Acta,* **655:** 243
Roychoudhury, R., Jay, E. and Wu, R. (1976) *Nucleic Acids Res.,* **3:** 101
Villa-Komaroff, L., Efstratiadis, A., Broome, S., Lovemico, P., Tizard, R., Naber, S. P., Chick, W. L. and Gilbert, W. (1978) *PNAS,* **75:** 3727

Chapter 20

Bahl, C. P., Marians, K. J., Wu, R., Stawinsky, J. and Narang, S. (1976) *Gene,* **1:** 81
Bahl, C. P., Wu, R., Brousseau, R., Sood, A. K., Hsiung, H. M. and Narang, S. (1978) *Biochem. Biophys. Res. Commun.,* **81:** 695
Ferreti, L. and Sgaramella, V. (1981) *Nucleic Acids Res.,* **9:** 85
Hoopes, B. C. and McClure, R. R. (1981) *Nucleic Acids Res.,* **9:** 5493

Chapter 21

Clarke, L. and Carbon, J. (1976) *Cell,* **9:** 91
Frischauf, A. M., Lehrach, H., Poustka, A. and Murray, N. (1983) *J. Mol. Biol.,* **170:** 827
Karn, J., Brenner, S., Barnett, L. and Cesarini, G. (1980) *PNAS,* **77:** 5172
Maniatis, T., Hardison, R. C., Lacy, E., Lauer, J., O'Connel, C., Quon, D., Sim, D. K. and Efstratiadis, A. (1978) *Cell,* **15:** 687
Tilghman, S. M., Tiemeier, D. C., Polsky, F., Edgell, M. H., Seidman, J. G., Leder, A., Enquist, L. W., Norman, B. and Leder, P. (1977) *PNAS,* **74:** 4406

Chapter 22

Gubler, U. and Hoffman, B. J. (1983) *Gene,* **25:** 263
Helfman, D. M., Feramisco, J. R., Fiddes, J. C., Thomas, G. P. and Hughes, S. H. (1983) *PNAS,* **80:** 31
Okayama, H. and Berg, P. (1982) *Mol. Cell Biol.,* **2:** 161
Okayama, H. and Berg, P. (1983) *Mol. Cell Biol.,* **3:** 280
Short, J. M., Fernandez, J. M., Sorge, J. A. and Huse, W. D. (1988) *Nucleic Acids Res.* **16:** 7583
Young, R. A. and Davis, R. W. (1983a) *PNAS,* **80:** 1194
Young, R. A. and Davis, R. W. (1983b) *Science,* **222:** 778

Chapter 23

Chia, W., Scott, M.R.D. and Rigby, P.W.J. (1982) *Nucleic Acids Res.,* **10:** 2503
Collins, J. and Brüning, H. J. (1978) *Gene,* **4:** 85
Collins, J. and Hohn, B. (1978) *PNAS,* **75:** 4242
Gitschier, J., Wood, W. I., Goralka, T. M., Wion, K. L., Chen, E. Y., Eaton, D. H., Vehar, G. A., Capon, D. J. and Lawn, R. M. (1984) *Nature,* **312:** 326
Ish-Horowitz, D. and Burke, J. F. (1981) *Nucleic Acids Res.,* **9:** 2989
Wolfe, J., Erikson, R. P., Rigby, P.W.J. and Goodfellow, P.N. (1984) *EMBO J.,* **3:** 1997

Index